W9-ATT-203

PRAISE FOR
FOR ALL HUMANKIND

"An absolute delight! By telling the story of the 1969 Apollo 11 moon landing through the eyes of observers from around the world, Dr. Tanya Harrison and Dr. Danny Bednar bring a freshness to it that is utterly beguiling. I would defy anyone not to be inspired by these extraordinary accounts from people who were, in turn, inspired by what they saw and experienced over fifty years ago. I know I was."

—Dr. Andrew Maynard, scientist and author of *Films from the Future* and *Future Rising*

"Harrison and Bednar's rich narrative serves to make the moon landing an inclusive event in human history. Told through a diverse set of characters from every continent, they deftly explore the intersectional impact of humankind's biggest step."

—Zara Stone, journalist and author of *The Future of Science Is Female*

"Tanya Harrison and Danny Bednar invite us to re-live one of humanity's proudest moments through a series of vivid, intimate, and refreshingly diverse accounts that challenge our perspectives and remind us that space exploration is a global pursuit with global benefits. Like the Apollo program itself, *For All Humankind* is both momentous and inspiring, the kind of stories that stay with you forever."

—Kellie Gerardi, Scientist-Astronaut candidate with Project PoSSUM

"A beautiful demonstration of how curiosity and wonder brought our planet together to accomplish the impossible."

—Dagogo Altraide, creator of ColdFusion and author of *New Thinking*

"The messages of hope and inspiration in this book are very much of a time, but they are also timeless. Maybe they speak to the ability humans have to overcome seemingly impossible challenges...but apparently only when we feel like it. As we stand at the new crossroads of space exploration and look forward, perhaps we should glance backward, too, and remember from whence we came. Because Apollo set the bar high. Very, very high."

—Geoff Notkin, president of the National Space Society and former host of *Meteorite Men* on The Discovery Channel

FOR ALL
HUMANKIND

FOR ALL HUMANKIND

✦

The Untold Stories of
How the Moon Landing
Inspired the World

TANYA HARRISON & DANNY BEDNAR

mango
PUBLISHING

CORAL GABLES

Published by Mango Publishing Group, a division of Mango Media Inc.

Cover, Layout & Design: Morgane Leoni
Illustrations page 33, 57, 76, 91, 117, 132, 146, 158: © Ray Brisendine

For permission requests, please contact the publisher at:
Mango Publishing Group
2850 S Douglas Road, 2nd Floor
Coral Gables, FL 33134 USA
info@mango.bz

For special orders, quantity sales, course adoptions and corporate sales, please email the publisher at sales@mango.bz. For trade and wholesale sales, please contact Ingram Publisher Services at customer.service@ingramcontent.com or +1.800.509.4887.

For All Humankind: The Untold Stories of How the Moon Landing Inspired the World

Library of Congress Cataloging-in-Publication number: 2019944129
ISBN: (print) 978-1-64250-096-7, (ebook) 978-1-64250-097-4
BISAC category code SCI098000—SCIENCE / Space Science

Printed in the United States of America

Danny: To my parents, siblings, and grandparents for starting my path. To Phil for inspiring me to make it my own. To Arnold for teaching me to value people and experiences along the way. To Elly for reminding me never to let hate chose the direction. And to Meoward for being by my side and not scratching the good couch.

Tanya: To Cassie Stuurman and Jenna Wakenight, for always being there.

CONTENTS

PREFACE

On July 20, 1969, American astronauts Neil Armstrong and Buzz Aldrin became the first humans to set foot on a celestial body beyond Earth, while Command Module Pilot Michael Collins kept dutiful watch from orbit overhead. Set against the backdrop of the Cold War and a "space race" between the United States and the Soviet Union, this amazing achievement could have easily been touted as a win for the US, laden with patriotic messages of America being the first to land people on the Moon.

Instead, alongside the two American astronauts, Apollo 11's *Eagle* lander brought messages and mementos of world peace to the lunar surface, including a stainless steel plaque that reads:

HERE MEN FROM THE PLANET EARTH
FIRST SET FOOT UPON THE MOON
JULY 1969, A. D.
WE CAME IN PEACE FOR ALL MANKIND.

Neil Armstrong also left a silicon disc on the Moon etched with microscopic goodwill messages from the leaders of seventy-three countries. Most of the messages were written in their respective native languages. In this era before computers were commonplace, some of these messages were handwritten, while others were typed. A great many of these messages call for world peace, pointing to the

✦ For All Humankind ✦

exploration of space as a chance for humanity to work toward something larger, together.

Even the Apollo 11 astronauts themselves recognized that their mission was far bigger. They chose not to have their names included on the mission patch—a break from NASA tradition. Michael Collins said that this choice was made so that it would be "representative of everyone who had worked toward a lunar landing."

The mission itself wasn't a solely American endeavor: A Canadian company built the legs for the *Eagle*, the two-person landing craft that carried Armstrong and Aldrin to Tranquility Base, their temporary home and landing site on the Moon. Australian radio telescopes received the live footage direct from the moon and relayed it to NASA's Mission Control in Houston for the world to see. German rocket engineers were critical in the design of the massive Saturn V rocket for the mission.

The global impact of the Moon landing was unquestionable. An estimated 600 million people worldwide watched the landing live—nearly one-fifth of the global population at the time. During the "Giant Leap" tour after returning from the Moon, during which the three Apollo 11 astronauts toured twenty-four countries in the span of thirty-seven days, over 100 million people gathered to see astronauts in person. Countries spanning every continent, from Belarus to Malawi, Mongolia to Australia, have released stamps commemorating Apollo 11 in some way. And yet, most of the stories we hear of the Apollo 11 moon landing in popular culture are American-centric.

I wanted to hear the stories of others, to get a worldwide perspective on this momentous event in human history. This book brings together the memories of the Moon landing from eight women and men who watched the event from outside the US, including a Lithuanian holocaust survivor, a Sudanese engineer, and an Indian primary school teacher. To fully capture the representation of humanity in this historic event, we made the decision to change the wording of this book's title from the original quote "for all mankind" to "for all humankind" so that everyone reading this will know that space is for them. Space is for *everyone*. We all belong to the universe, and together we can all be awed and inspired by what is possible.

Tanya Harrison

INTRODUCTION TO THE
APOLLO PROGRAM

In this book you're going to read eight different stories about the Apollo 11 mission. These stories were told to us by people from all around the world, who experienced Apollo 11 in very different ways. But before going back in time and across five different continents, we thought it would be helpful to provide some information about NASA's Moon landing program, Apollo.

THE "SPACE RACE"

Humanity's course toward the Moon started in October 1957 when Russian scientists launched a rocket into space carrying a small metal sphere about the size of a beach ball. This rocket was traveling fast enough that when the metal ball was released from the nose cone, it went into orbit around the Earth. This metal ball was called Sputnik—humanity's first artificial satellite. By most accounts, Sputnik officially kicked off the "Space Race."

The term "Space Race" is used to describe a sort of technology Olympics between two countries: The United States of America and Russia—more often referred to at the time as the Union of Soviet Socialist Republics (USSR), or the Soviet Union for short. If we're keeping score then, it was the Russians who were leading the medal rankings in the early days of the "Space Race".

On April 12, 1961, the Russians scored their second big win. A young man named Yuri Gagarin sat cramped in a tiny cockpit atop a Russian Vostok rocket. Gagarin was a military pilot, but his job that day was to be a passenger. The rocket on which he sat was significantly bigger than the one used to launch Sputnik four years earlier and didn't include any manual controls for Yuri to pilot the craft himself. After launching from Baikonur (in what is now Kazakhstan), Gagarin's rocket pierced through the atmosphere and entered space. Like Sputnik, the speed of the rocket propelled Yuri's cramped capsule into Earth's orbit—where no human had ventured before.

After just over an hour circling the Earth, Gagarin's Vostok capsule re-entered the atmosphere, deployed its parachute, and successfully touched down. His mission was a success, making him the world's first cosmonaut, the Russian word for astronaut (both "cosmonaut" and "astronaut" come from the Latin words for "star sailor").

Seemingly losing the "Space Race" at that point—or at least convincing themselves it was a race and that they were losing—American politicians and scientists set out to do something big. In May of 1961, President John F. Kennedy told the world that America was planning to put humans on the Moon. This was, to say the least, a bit of a shocker. The President was announcing this Moon plan mere weeks after the first human had even ventured into space. Few people before this had really taken a human Moon landing seriously. It seemed to many an impossible task. But Kennedy was confident—so much so that he gave America a deadline, saying, "I believe that this nation should commit itself to achieving the goal, before this decade is out, of landing a man on the Moon and returning him safely to the Earth."

Spurred on by the success of Sputnik and Yuri Gagarin, and against a looming deadline, engineers and scientists from around the world were brought to work at NASA, the National Aeronautics and Space Administration. Far from the shining symbol of excellence it is viewed as today, NASA was barely a toddler when Kennedy made his announcement. It had only been created three years earlier.

NASA's attempt to put humans on the Moon would be called Project Apollo, and it had about nine years to meet the President's deadline. This gave those scientists and engineers less than a decade to do something people had barely started thinking about from a technical perspective. The clock was ticking.

SO, YOU WANT TO LAND ON THE MOON?

If you want to land on the Moon—and come back to Earth—you're going to need at least two things: a gigantic rocket, and a spacecraft able to land and come back.

At this point it's important to note that spaceships in sci-fi movies are very different from spacecraft that were built in the 1960s, or even those built today. In movies, spaceships are usually an all-in-one vehicle, like the Millennium Falcon or the starship Enterprise. Fictional ships like this take off from planets, fly around in space, and land all on their own, all in one piece. In real life, we aren't quite there yet.

It's incredibly hard to get off of Earth. Our planet has a lot of gravity that tries to pull you right back down. Long story short, you need a very powerful rocket to leave the Earth's gravity. But once you're in space, the rocket has done its job and there's no further need to lug it around with you. We don't need the rocket to get back down to Earth either. That's where gravity is our friend. We can just fall back to Earth, albeit in a controlled manner so as not to burn up in the atmosphere.

So, what we usually do is build small spacecraft called capsules, and put them on top of a powerful rocket. The rocket lifts the capsule into space, where it then separates and goes wherever it needs to. The rocket, meanwhile, falls back down to Earth.

For the Apollo program, NASA put three astronauts in a capsule, put the capsule on top of a rocket, and launched them toward the Moon. This sounds simple enough, but there was one minor detail to attend to: No one had ever built a rocket that was able to do this. No one had ever built a rocket that could lift a spaceship and three people at a fast enough speed that they could escape the gravity of Earth and get to the Moon.

To add to NASA's problems, no one had ever designed a spacecraft able to land on another world and take off again. Both NASA and Russia had sent robotic missions that landed on the Moon in the late 1960s, but none of these were designed to return to Earth or carry humans.

After exploring all the options for getting to the Moon, landing, and getting home, NASA came to a conclusion: they needed one very big rocket and two different spacecraft.

The plan went like this: The massive rocket would have the two spacecraft placed on top of it. There would be three astronauts onboard, right at the top. The giant rocket would push both spacecraft into space. Once in space, while still near the Earth, the two craft would connect and start flying towards the Moon. The rocket would then fall back down to Earth having completed its job.

Once the two spacecraft got to the Moon a few days later, they would detach from each other. One would remain flying above the Moon (staying in orbit) with one astronaut onboard. The other craft, carrying two astronauts, would land on the Moon's surface.

The astronauts that made the journey down to the lunar surface would walk around for a bit, collect some rocks, then launch off the Moon back up to the other spacecraft that had remained in orbit. Thankfully, it's much easier to launch off the Moon than the Earth because there is no atmosphere, as well as much less gravity holding you down. Getting off the Moon only requires a small booster.

After getting off the Moon and reaching the other craft, the two would reconnect and the Moonwalking astronauts would rejoin the third that had stayed in orbit above them. Reunited in the same spacecraft, the three astronauts would make their way back home to Earth. Since they didn't need two spacecraft to go back home, they would leave one to fall back down to the Moon just before they left.

As our three astronauts returned to Earth, the Command Module would fly through the atmosphere and deploy multiple huge parachutes—each one nearly half the size of an Olympic swimming pool—and land softly on the ocean. They would float there until being picked up by a naval carrier and brought back to shore.

No one said going to the Moon would be easy.

THAT BIG ROCKET: THE SATURN V

Next time you're in a city, find a building that's about thirty stories high. That's the height of the rocket built by NASA to send humans to the Moon. The rocket was called the Saturn Five. The "five" is always written in Roman numerals, so it's written out as "Saturn V." To this day, over fifty years later, the Saturn V is still the most powerful vehicle ever built by humans.

Like a lot of rockets, the Saturn V was what's called a multistage rocket. These rockets shed weight as they fly. This is good as lighter things can fly faster and will have an easier time escaping the pull of Earth's gravity. Multistage rockets are built in sections ("stages") and burn fuel from the bottom up. When the fuel from one section is used up, that section breaks off and the rocket becomes lighter.

Spacecraft 1: The Command/Service Module

The Command/Service Module was a three-seater capsule that sat right at the top of the Saturn V. When the Saturn V was on the launch pad, the astronauts would be sitting with their backs to the ground looking straight up at the sky. The only thing on top of this craft was a small escape rocket for emergencies. This escape rocket was like an ejection seat for all three astronauts, but it was only needed in case of an accident on the launch pad; it would be discarded right after takeoff if unused and wouldn't go into space.

The Command/Service Module had two parts: the Command Module and the Service Module, hence the name "Command/Service Module." The Command Module was a triangular capsule at the front where the astronauts sat and controlled everything. The Service Module was the cargo part of the spacecraft behind them. NASA figured it would take the

+ For All Humankind +

astronauts about three days to get to the moon. This meant they would need about seven days worth of food, power, and oxygen to cover the day at the moon and the return trip. Essentially, the Service Module was the supply closet for the mission. To simplify things, from here on out, we'll just call this whole craft the "Command Module."

NASA also attached a very small rocket engine to the back of the Command Module so that the astronauts could make maneuvers on their way to the Moon. This was critical for entering the Moon's orbit, which required precise flying to account for the Moon's own, albeit weaker, gravitational pull. When an Apollo mission was over, the Command Module would return to Earth's orbit from the Moon and split in two. Only the triangular "command" part with the three astronauts onboard would return safely to Earth's surface. The supply closet that was the Service Module was left to orbit the Earth and eventually crash into the ocean.

Spacecraft 2: The Lunar Module

At this point we've got a big rocket, a place to sit and control the spacecraft, and a place to keep all the important things like food and oxygen. So, what else do we need? Well, a huge part of going to the Moon was flying down to the lunar surface, landing, and being able to come back.

For this part, NASA had some decisions to make. They realized that the Command Module couldn't perform the actual moon landing once it got there. Instead, the solution was to build an entirely separate spacecraft that came along for the ride and would only be used for going up and down to the lunar surface. This craft would attach to the front of the Command Module for the trip to the Moon, kind of like driving to a lake with a canoe on the roof of your car. The second spacecraft was designed specifically to land on the moon, and was called the Lunar Excursion Module, or LEM for short (pronounced like the beginning of "lemon").

The LEM looked a bit like a robotic spider. It was designed only for landing on the Moon and getting back off again. After landing, the LEM would be the astronauts' home base—their tent in the woods, so to speak. When they had their spacesuits on and were ready, the astronauts would go outside and explore the lunar surface. In the case of Apollo 11, the two astronauts explored for about three hours before getting back in, taking a nap, and flying back up to their astronaut colleague in orbit. In order to do this, the LEM would actually split in half, leaving its legs on the Moon's surface. The top half, lifted by a small booster engine, would then rendezvous with the Command Module orbiting in wait above them.

Once reconnected to the Command Module, the astronauts would unload the Moon rocks they collected and jump back in the Command Module. Just before heading back to Earth, they would detach what was left of the LEM and let it crash into the Moon, like leaving your canoe to drift on the lake if you don't want to bring it all the way back with you. (Please don't actually do that with your canoe though. Littering isn't cool.)

THE ASTRONAUTS

Who would NASA need to take on this mission? They decided that each Apollo mission would need three astronauts: one person to fly the Command Module, one person to fly the LEM, and an overall mission commander. The Command Module pilot would be the "unlucky" one who didn't get to go down to the Moon. Instead, they would stay in orbit flying above the Moon while the other two went down in the LEM. The first Moon landing was accomplished by three astronauts on the Apollo 11 mission: Neil Armstrong, Buzz Aldrin, and Michael Collins. You'll hear a bit more about them throughout the rest of the book.

Practicing for the Moon

The success of Apollo 11 didn't come out of nowhere. Just like learning to do anything difficult, NASA took incremental steps in achieving the Moon landing. This involved a series of missions, each testing out the equipment, preparing the astronauts, and making sure they were ready for everything and anything that might come their way.

Sadly, during this testing there were casualties. Apollo 1 was meant to be a test of the Command Service Module in low Earth orbit. It had a crew of three astronauts: Virgil "Gus" Grissom, Ed White, and Roger Chaffee. Tragically, a fire in the module during launch rehearsals on the ground resulted in the deaths of all three astronauts. This mission had originally been designated AS-204, but the widows of the three astronauts asked that the mission be called Apollo 1 in their honor.

Two launch tests of the Command Module had taken place without a crew aboard previously, so the naming scheme for the next mission jumped to Apollo 4; there was never technically an Apollo 2 or Apollo 3 mission. Apollo 4 and 6 marked the first and second flights, respectively, of the mighty Saturn V rocket, still with no one onboard. Apollo 5 was the first uncrewed test flight of the LEM using the Saturn V's little sibling, the Saturn IB. All of these tests were to make sure the equipment would be safe for humans before taking the risk of actually putting astronauts into the mix.

The first Apollo astronauts to fly were the Apollo 7 crew; Walter Cunningham, Walter Schirra, and Donn Eisele. Apollo 7's goal was to test out the Command Module. They did this by flying it around the Earth a few times and making sure everything worked. The astronauts on the Apollo 8 mission had the same goal of testing the Command Module, but they were lucky enough to be asked to test it by flying around the Moon. On this mission, Frank Borman, Bill Anders, and Jim Lovell became the first humans to truly leave Earth and its orbit. Apollo 9, with James McDivitt, Dave Scott, and Rusty Schweickart, followed by staying back in Earth's orbit and testing both the LEM and Command Module together. This was important to make sure the two spaceships could be connected and disconnected when needed.

Apollo 10 combined everything that had happened so far, flying both the Command Module and LEM to orbit around the Moon. Astronauts Gene Cernan, Tom Stafford, and John Young even detached the LEM from the Command Module and flew it within fifteen kilometers of the lunar surface—practicing almost everything needed to land on the Moon except actually landing. Because of this, Apollo 10 is often referred to as the dress rehearsal. In fact, NASA didn't even include enough fuel for the mission to be able to land and take off again as it wasn't intended to be done on this mission. Apollo 10 astronaut Eugene Cernan later joked that NASA might have done this intentionally so that he and Stafford couldn't take it upon themselves to just go ahead and land on the Moon.

By 1969, NASA had gone step by step and were ready to attempt a lunar landing. They had tested the Command Module above Earth (Apollo 7) then above the Moon (Apollo 8). They had tested the Command Module and LEM together above the Earth (Apollo 9) and then, again, above the Moon (Apollo 10). Everything had been done at least once...except the actual landing. That would be left for Apollo 11 to try for the first time. You'll read all about Apollo 11 in the rest of this book, so we won't talk about it here.

Six more Apollo missions flew after Apollo 11. Five of these were successful in landing on the Moon: Apollo 12, Apollo 14, Apollo 15, Apollo 16, and the finale, Apollo 17. One other mission launched, but the astronauts had to fly around the moon without landing because of a malfunction in the Service Module. This was the famous Apollo 13 mission.

What all these missions had in common was the goal of landing on the Moon and, most importantly, getting back home safely again. Amazingly, each Apollo mission that launched accomplished the goal of getting the astronauts home safely.

A TRULY GLOBAL MISSION

To this day, the Saturn V remains one of the most complex machines humans have ever built. When it first launched from Cape Kennedy In 1967, now known as Cape Canaveral, NASA's Moon rocket was the culmination of centuries of human ingenuity. The first rockets were invented in China and India almost one thousand years earlier. The mathematical formulas necessary to calculate the rocket's flight and that of the Apollo 11 spacecraft used algebra developed by ancient Babylonians and perfected by Persian scholars in the ninth century. Work by seventeenth-century German astronomer Johannes Kepler and British physicist Sir Isaac Newton provided many of the basic equations for the orbital mechanics needed to get Apollo 11 on and off the Moon. And in the decades leading up to Sputnik, engineers from around the world studied the work of pioneering Russian rocket scientists such as Konstantin Tsiolkovsky, who developed the crucial equations necessary for rockets to reach outer space.

In short, it took all of humanity's ingenuity to launch the Saturn V.

And it wasn't just the science and technology of past generations taking part. Along the way, people from around the world were with Apollo 11, even inside their spacecraft.

Among the personal items Armstrong and Aldrin brought with them to the Moon was a cassette of Antonin Dvořák's 9th Symphony. This nineteenth-century masterpiece of classical music, written by the Czech composer while he lived in the United States, was inspired by African American folk music. Dvorak wanted a symphony that expressed the musical potential of Indigenous American and African American cultures, which for too long had been ignored by other composers. The result: a blend of African American folk melodies and Slavic compositional traditions wrapped into a single musical masterpiece—a masterpiece that found its way to the Moon in July 1969.

What you will see in the pages to follow is that the Apollo missions were some of the most incredible adventures humanity has ever witnessed, and just as the world inspired Apollo, the mission, through its accomplishments, returned the favor. Throughout the 1960s, people from all around the globe watched these missions as they happened live, gripped with excitement and wonder.

Apollo didn't happen in a vacuum (pardon the pun). It was intertwined with millions of people's lives, as participants or observers. In this book, we want you to feel what it was like to be on Earth when humans first touched another world. We also want you to get a sense of what life was like in this different time, five decades ago. From India to Canada,

Sudan to Iran, we want to take you around the world and back in time.

We hope you enjoy this sort of global folk history of Apollo that we've had the pleasure of hearing and assembling. We are honored to share these stories with the world and, most of all, want to thank those who shared their precious time and memories with us. They have provided us all with a gift, to have the opportunity to relive this unique time in human history.

CHAPTER 1

Elly, Canada

Lying in bed staring up at the stone ceiling of the St. Ottilien Monastery, fatigued, exhausted, and still recovering from the most horrible of atrocities, Elly was torn between emotions. A part of him was relieved and felt safe for the first time in years. The other part of him was intensely angry and overcome with hatred.

It was the spring of 1945. Elly Gotz was seventeen and recuperating in a makeshift hospital near the town of Geltendorf in the south of Germany. When he looked up from his bed, he saw a three hundred-year-old stone ceiling held up by old wooden arches. The monastery was a cold place of mostly hard surfaces like stone floors and wood benches. The wheeled-in hospital cots were the only soft surfaces, and the first comfort Elly and most other people there had felt in years.

The long halls of the monastery housed dozens of other weakened men, women, and children. All had just been rescued from the main Nazi concentration camp at Dachau, Germany. Among these men was Elly's father, a fifty-four-year-old Lithuanian who now barely weighed his own age in pounds. Elly himself only weighed about seventy pounds and struggled to lift his emaciated body from the bed. When he did have the strength, he would walk the stone stairs of the monastery to exercise and rebuild what muscle he could.

If he wasn't exploring the grassy surroundings of the monastery, Elly was with his father and the other survivors in their room. Conversations filled the long stone halls with lively voices: German, Polish, Hebrew, Yiddish, and Russian, among others. The rows of beds the only soft surfaces to mute the echoes.

Like everyone there, Elly had been rescued from the darkest of situations. His body saved from slave labour and starving conditions that would have taken his life eventually, his mind saved from the constant presence of death and the thought that, at any moment, he or his father could be killed.

While the physical torture of the Holocaust was over, Elly's mind remained in a dark place. He was full of hate for those who had done such horrible things to him and millions of other Jewish people across Europe. By the end of World War II, over fifteen million people had been killed by the Nazi regime. Six million of these were European Jews, singled out for genocide by the Nazis and murdered in mass killings at concentration camps as part of the Holocaust—what the Jewish community now calls the Shoah.

Slowly gaining back their strength, Elly and his father spent their days trying to find Elly's mother and the rest of the family, sending frantic letters to the Red Cross. As they did this, Elly wondered: Could he ever live a meaningful life? Was he to be forever full of hate for the people who had done this? At night, when he couldn't sleep, he thought about the hate that had begun to overwhelm him. He knew his entire life would be defined not only by what happened to him and his family during the war, but also by how he chose to let it affect his future. He had a choice to make.

•••

Four years later, Elly was in Johannesburg, South Africa, studying to become an electrical engineer at Witwatersrand (Wits) University. To do so he had to leave his parents who were living in Rhodesia (known today as Zimbabwe). Despite the difficulty of being away from his family after they had endured so much to be reunited, he was thrilled to be attending university and learning about two of his favorite subjects: physics and engineering.

For Elly, the best part of university was being surrounded by other people interested in science. He and his friends would come to class early to talk about everything happening in the world. At the time, major scientific fields were undergoing a revolution. Albert Einstein's theories were still new, being hotly tested and debated in physics lectures around the world. The computer chip had only recently been invented and was paving the way forward beyond the primitive technology of World War II, and engineers in America and Russia were successfully launching rockets high enough to pierce the atmosphere and into space.

Elly would walk to class among Johannesburg's world-famous Jacaranda trees that gave the Wits University campus a bright purple hue amongst the otherwise green landscape. As beautiful as the Wits campus was, his gaze was never limited to just the greenery. Whenever a plane flew overhead, Elly would stop. He would try to see if he could tell what kind it was. He was absolutely fascinated by flight. Before the war, his dream had been to become a pilot, to control a machine in the air and leave the ground behind.

After the war, things felt a bit different. Becoming a pilot didn't seem realistic. First and foremost, Elly wanted an education. He decided to become an engineer, and his talents made him well-suited to chase this goal. He loved to create new ways of doing things, to invent machines to solve problems, and to question old beliefs. In short, he was a born inventor, able to think differently than others his age and always curious if things could be done a new way...though this habit of questioning things didn't always do Elly favors.

One semester, when his class was learning about the makeup of waveforms, Elly doubted his professor's claim that an irregular wave can be broken into an endless number of regular sine waves. The professor proceeded to schedule a lesson specifically to prove that the theory was correct. That day Elly happened to be late, arriving a few minutes into the lecture. As he walked into the lab, he immediately heard the professor say, "There he is, the doubting Thomas"—an old nickname for people who are skeptical by nature.

After the professor successfully proved the nature of irregular waves, Elly would have to put up with being called "doubting Thomas" for a few more years. But he wasn't bothered by his reputation for being curious and questioning. He knew his curiosity drove his desire and ability to learn. Eventually it paid off. In 1952, Elly graduated from Wits University with a degree in electrical engineering.

After graduation he moved back to Rhodesia to reunite with his parents. At that time, the economy there was far from booming. Unable to find a job as an engineer, Elly put his degree to work and opened a radio repair shop. His people skills and technical know-how led to success, and he eventually opened a recording studio. In these years Elly met his wife, Esme, and had three children. When the opportunity presented itself, he moved back to South Africa to open and operate a plastics factory.

Sadly, South Africa in the 1950s brought new scenes of horror to Elly's life. Racism was everywhere. Many of the white people of South Africa treated black Africans as second class, lesser humans. Elly had seen it before. He worried what would happen to his children if his family remained there, not wanting them to grow up surrounded by racism and the violence that follows it.

+ For All Humankind +

Because of the atmosphere in South Africa, Elly and his wife knew it was time to leave. Nothing good could come from being surrounded by so much hate, and as a recent immigrant Elly was powerless to stop it. In search of a better life for their children, he and his wife moved their family to Toronto, Canada.

They arrived in Toronto in 1964, where they bought a bungalow in the suburbs. Elly joined his brothers-in-law in operating a plastics factory in the industrial heart of the city. Canada was different from the other places he had lived, and the adjustment for his family was at times a challenge. But moving to North America came with perks as well. Elly noticed that the Canadian newspapers were full of news about NASA and spaceflight.

The engineer in him wanted to know everything—how NASA built the spacecraft, how they designed the fuel pumps, how they balanced such huge machines...everything. At the same time, Elly had never let go of his childhood dream. The aspiring pilot in him wanted to know what it felt like to sit on a rocket and launch into outer space.

In time, Elly became very successful in Toronto. His drive to learn, invent, and try new things soon paid off, presenting him with rewarding experiences and allowing him to support hundreds of people by providing them with well-paying jobs at his factory. He was also able to save for his children's education, knowing they wouldn't have the same difficulty he did getting high school diplomas or going to college.

Eventually, Elly and his family became Canadian citizens. He had lived in Lithuania, Germany, Norway, Rhodesia, South Africa, and now Canada. So, while he was now a proud Canadian, Elly Gotz was truly a citizen of the world. Ultimately, though, what mattered to him was not national identity or pride, but that he and his family were safe and happy. For this, Elly loved Canada. He became an active community member, volunteer, and a member of his synagogue. In the decade that followed, as a generation matured that had never experienced the Holocaust (including his own children), Elly began speaking publicly about the horrors of the Holocaust. Through his own story he would impress upon countless children the danger of one simple feeling: hate.

•••

It was a hot Sunday in July. Elly's home was empty. The kids were at their cottage by the lake with Esme. He spent most of the day doing a mix of reading, fixing things around the house, and working in the yard. It was an incredibly warm day in Toronto—with the humidity it felt nearly 30°C (about 86°F) all afternoon.

After a few hours in the garden, it was about four o'clock. The summer sun was unrelenting, even as its angle was starting to allow for more shade behind the house. It was time for a break. To cool himself down during the day, Elly would make his way into the basement. While the basement was significantly cooler than the rest of the house, there was another reason he wanted to go down there.

Today was the day that NASA would be landing on the Moon, and the family's brand new color TV was in the basement. Elly needed to keep checking in to see how the mission was progressing. He sat on the green basement couch and put his glass of water on the wooden coffee table in front of him. As condensation quickly formed around the outside of the glass, he turned on the TV.

The TV's sound came on before the cathode ray screen had time to warm up and show a picture. Elly could hear newscasters describe the status of the mission. When the screen turned on, he saw that they were describing the landing procedure over images of NASA's mission control room in Houston. He had gone downstairs just in time. The lunar landing was imminent.

Neil Armstrong and Buzz Aldrin were on a descending flight path, taking them closer and closer to the lunar surface, with each kilometer they traveled like a plane coming in for landing. They would soon land at their target in the Mare

Tranquillitatis region, just above the Moon's equator on the side facing Earth (the same side of the moon always faces Earth).

In that moment, Elly actually began to feel a bit nervous. With a background in electrical engineering, he knew how many points of failure there were in a machine as complex as the Lunar Module. Just one circuit failure or blown transistor and the astronauts might not be able to land. The entire event would be a disaster and two men would be marooned to die in space.

With the sun's beams still blasting through the narrow basement windows, Elly listened to the voice of the Canadian Broadcasting Corporation (CBC) news anchor explaining that there would be no video of the actual landing. The TV cameras could only be turned on and connected to Earth once they were safely on the moon's surface. The screen cut to a countdown clock and animation of the Lunar Module. Elly leaned forward and listened closely, cautiously, as the live audio from the astronauts played over a cartoon animation of the spacecraft descending.

He could hear Buzz Aldrin speaking with ground control in Houston, as well as to Armstrong, who was piloting the LEM. Buzz was reading out numbers from the computer, rattling off how high they still were above the surface and how much fuel was remaining. The power of those computers

Buzz was reading from was unimaginable to Elly—they were the best of the best and were designed specifically to help humans fly in this strange environment on the Moon. He grew jumpier as he started thinking about the hundreds of transistors and circuits that must be operating at full capacity. After all, in his time running factories and radio repair stores, he had seen hundreds, if not thousands, of blown fuses, corroded connections, and faulty components. Knowing the nature of machines, he couldn't help but be tense. He grabbed his glass of water, the ice cubes mostly melted away by now, and took another sip, hoping for the best.

As nervous as he was, the aspiring pilot and adventurer was still alive in Elly. This side of him wasn't nervous at all, but actually jealous. What did it feel like to see the Moon's surface coming increasingly closer to you? How did it feel to fly with no atmosphere? With no resistance, did it feel like the spacecraft wanted to drop like a stone? Could the astronauts feel the thrust of the descent engine? Every push of the navigation thrusters? What would they do if the main descent engine failed? The Lunar Module was like a single engine aircraft, with no backup plan if things went wrong. If an airplane experiences engine failure on Earth, pilots can glide to a safe landing thanks to atmospheric lift. But the moon has no atmosphere—if the engine failed, the

Lunar Module would immediately plummet to the surface and crash.

Just as he continued to think about all of the possible electrical risks, mechanical risks, and flight risks, Elly heard Neil Armstrong interrupt a brief period of silence: "Houston, Tranquility Base here. The *Eagle* has landed."

They had done it.

Finally able to relax, he leaned back into the couch and smiled. Forty-one years old, father of three, and a citizen of the world, Elly Gotz had one simple reaction: "Wow, what a moment!" Two astronauts had piloted a spacecraft to the surface of the Moon while he listened live back on Earth.

It was about 4:15 in the afternoon. The CBC newscaster informed the audience that it would be another few hours until Neil and Buzz exited the Lunar Module to explore the surface. Elly watched for a few minutes more, then finished his water, which had no ice cubes remaining, and went back upstairs.

He'd initially had planned to do a few more things around the yard, but admitted defeat when he realized he was simply too excited. It was a beautiful day, and the most exciting thing he had ever seen on TV had just happened. The kitchen upstairs was lit up with sunlight, the shine reflecting Elly's mood perfectly.

Arguably, the most famous image of the twentieth century. This picture of Buzz Aldrin was taken by Neil Armstrong, who played photographer for much of the Apollo 11 mission.

Deciding he was indeed far too excited to go pull weeds from the garden or take on any other type of chore, he instead picked up the phone and started calling family members. He was ecstatic, but had no one to talk to in an empty house. He absolutely had to talk to somebody about what had just happened. He wanted to share what he had just seen, to talk about how amazing it was, and, ever the engineer, to make sure people appreciated the magnificence of the machinery.

Elly had spent years fixing radios and all sorts of other electronic devices. While radios are complex machines, they were nothing compared to what NASA had built to go to the Moon. Maybe, he thought, people who aren't engineers can't fully appreciate the technological immensity of what NASA engineers accomplished today. A machine can be very stubborn when it doesn't want to work. Did people who didn't work with machines understand how incredible it was to use them to land on the Moon? It takes great engineers to design a machine that absolutely cannot fail during its one and only opportunity to work.

Calling his friends and family, his voice was beaming with pride and excitement. Elly made sure to emphasize the amazing nature of the landing, insisting people appreciate the mastery of machines that NASA had just shown. For the next few hours he regaled as many people as he could over the phone with everything he knew about the space program (and engineering).

As evening approached, it was starting to cool off outside. The humidity was relenting too, at least a bit, and the sun was past the horizon. Only a slight bit of light still pierced the air. It was setting up to be a calm, warm, summer night. Having made his last phone call, Elly got ready to watch more of the mission. He went to the kitchen to make a sandwich, pour another glass of water, and look through the pantry for some snacks to satisfy his sweet tooth. Sadly,

+ For All Humankind +

most of the good stuff had been packed up with the kids to go to the cottage. But, he finally spotted potato chips and chocolates—the perfect snack foods to relax downstairs with and watch history be made.

With no wind or rain outside, the streets were quiet on this Sunday night. The only sound in the house came from the TV. Elly wondered: was everyone else inside doing the same thing he was? He could see the Moon through the short basement windows above him. It wasn't a Full Moon, barely at Third Quarter phase, but it seemed particularly bright. Inside, he sat on the couch and ate his sandwich as he listened to the newscasters of the CBC discuss what kind of preparations the astronauts were undertaking. By now it was nearly eleven o'clock at night.

Then, an image came on the TV. It was the side of the lunar lander. Elly had seen the craft dozens of times in the newspaper and during the news coverage of the Apollo 9 and 10 missions. But now it was sitting on the ground, a vantage point of the craft he hadn't seen before. That was because this was a live feed of the lunar module on the surface of the Moon. The news anchor confirmed what he was seeing. A small camera connected to a long arm on the side of the lander had been extended and turned on. The Lunar Module wasn't alone. Viewers could see Apollo 11 Commander Neil Armstrong holding onto the ladder along the side of the spacecraft, ready to step down at any moment. The image wasn't very good, even for TV standards at the time—it was black and white and seemed faint. But that didn't matter. Live video from the Moon was beaming into Elly's basement. That feat alone, being able to transmit a live signal to Earth from the surface of the Moon, was surreal in the moment.

After a brief conversation between Armstrong and Houston, which viewers were allowed to listen in on, the young man from Ohio, two years younger than Elly, began making his way further down the lander. Armstrong was stepping feetfirst down the ladder that connected the Lunar Module crew cabin to the ground. He even tested being able to jump back up the ladder.

Ever the engineer, Elly was worrying again.

There he sat in the basement of his family home, surrounded by pictures of his loved ones on the walls and the serene silence of the suburban neighborhood outside. He was nearly 400,000 kilometers away from these men on the moon. They had traveled there on the biggest rocket ever built, and landed using the most advanced computers ever created. All of it had gone according to plan, but Elly worried about one last part.

While living in South Africa, he operated a sound-recording studio. He knew the task of sending a television transmission from the moon must have been incredibly complex. The transistors, the power source for the reception dishes, the wiring in the camera...his mind never left engineer mode. What a terrible shame it would be if they went all the way there, accomplished this incredible feat, and nobody back on Earth got to see it because of a blown fuse.

All these worries immediately washed away as Elly saw that the television feed only needed to hold on for a few seconds more. "I'm going to step off the LEM now." Armstrong was standing on the base of one of the Lunar Module's feet and was about to put the first human footprint on another world.

"That's one small step for man, one giant leap for mankind."

Looking at a person standing on the Moon, he thought: "Beautiful." That is all Elly thought. "Beautiful."

For Elly, Armstrong's words could not have been more perfect. This was an accomplishment for all humankind. We had done it. We had set a seemingly impossible goal, filled with obstacles of every kind, and we had succeeded. There were scientific questions about the lunar environment, engineering challenges around how to build the machines, and innumerable human factors to overcome. Who could fly these machines? Was it worth the risk? Could three astronauts function properly for the duration of the mission? All of these challenges—scientific, engineering, and human—didn't matter now to Elly or anyone else. Humanity had just made one giant leap.

•••

The atmosphere the next day on the floor of Elly's plastics factory felt like he was back on campus with his university friends. Everyone was talking about what had happened on TV the night before. Elly in particular was singing the praises of NASA all over again and talking to everyone about the amazing event. He went around asking every worker in the building if they had seen it.

One employee replied, "Did I see what?"

"They went to the Moon in a giant rocket!" Elly responded in shock.

Apparently he still had some people to convert when it came to admiring spaceflight.

Of course, when Elly's family returned from the cottage, he had a new audience to tell all about NASA's Moon landing. They were used to it by now. When he was reading the newspaper every morning, he would often talk at length about the incredible machinery involved in landing on the Moon. And as a Holocaust survivor, he was also always quick to point out to his wife any time the news mentioned chief US rocket engineer Wernher von Braun, whom the American government had snapped up from Germany after the war. "Look! The Nazis terrorized poor London for years with their rockets, now von Braun is helping the Americans go to the moon!" Elly knew all too well that von Braun had used slave labor at concentration camps to build his rockets during the war; von Braun was the only part of the Space Race he didn't like reading about.

Over time, Elly's family would grow accustomed to his fascination with machines and flight. They had to. Fulfilling a childhood dream, he eventually got his pilot's license and his own plane. He flew his family around in a small single-engine airplane, taking them on vacations across Canada and the United States. While flying a small aircraft is always dangerous, the family trusted Elly. They knew he thought through every single risk and was prepared to deal with each one, mechanical or human. Calculating risk and knowing the details of his machines were things Elly always took seriously.

His understanding of both risk and machines certainly paid off. On one particular family vacation to the east coast of Canada, he landed with his family in Moncton, New Brunswick, to spend the day. After walking through the streets and sightseeing, he prepared to fly the family to the next stop for the night. There was a chance for severe weather, but it wasn't meant to arrive for several hours, and

Elly figured he could fly low to avoid it in the unlikely event that the storm moved in earlier than forecasted.

Shortly after taking off from Moncton airport, Elly couldn't see a thing. Rain covered his windshield. The clouds had moved in rapidly—three hours earlier than forecasted. The conditions were what pilots called instrument weather, where you fly entirely based on the readouts of the plane's instruments. Unfortunately, Elly didn't have an instrument license, and his family was onboard. It was a horrifying scenario.

But Elly kept calm. He knew his plane, and he knew he couldn't panic. He thought to himself, *if you panic now, you're all dead*.

Elly contacted the Moncton airport tower by radio and requested help to return to the runway. Air traffic control directed him to turn a precise number of degrees using his compass. Elly had no problem with the directions, he was intimately familiar with his airplane's instruments. From there they coached him to about 3,000 feet altitude and told him to fly level. He was only to descend when they said so. But even at that altitude, he still couldn't see a thing.

After several minutes of nerve-racking flight through dense clouds and rain, he heard the voice over the radio tell him to begin to descend at a particular rate. As he emerged beneath the blanket of clouds, through the continuing rain on the windshield, he could see the runway lights. He was right on track for a perfect landing.

From this experience, Elly learned that he didn't panic in dangerous situations. He would have made a fine astronaut. With some encouragement from Esme, Elly decided to get his instrument rating.

•••

Even fifty years later, Elly still sometimes looks up at the moon and smiles, thinking, *We've been up there!*

In so many ways, his life has paralleled the space age. When he was born, the earliest liquid fuel rockets—tiny versions of what would one day go to the Moon—were just being invented. When he graduated from university, he was learning about the first rockets to fly beyond Earth's atmosphere. When he moved to Canada, humans were landing on the Moon. And as he celebrated his ninetieth birthday, he was reading online articles about amazing little rovers on the surface of Mars. He even got a little bit closer to becoming an astronaut when, at the age of ninety, he jumped out of an airplane (with a parachute) to celebrate the 150th anniversary of Canada, his adopted home. Falling through the sky, he proved you're never too old to explore and take risks.

Elly has seen the worst of humanity, and, in some ways, the best. He is constantly reminded of how ingenious humanity can be: landing rovers on Mars, sending spacecraft to Pluto, and orbiting the Earth every day in the International Space Station—all wonders of human ingenuity. However, Elly also reminds himself, and others, how stupid we can be when we focus on the wrong things. When we choose the wrong emotions to motivate us. When we hate.

Elly knows firsthand how hate drives so many of our bad decisions. Hate blinds us. Hate makes us stupid. No one has ever accomplished important things fueled by hate. No one has landed a rover on Mars fueled by hate. No one has lived in space for a year fueled by hate. Hate is worthless. All the good of humankind, all the accomplishments of science and engineering throughout history, came from people who turned away from animosity. When Elly Gotz saw humans walk on the Moon, it was about more than the technology. It was about what getting there meant to humanity.

Two decades after the First World War, humanity started a second. But two decades after the Second World War, there was no third. Despite the fact that the use of weapons and technologies from World War II could have easily led to another horrendous global conflict. Thankfully, instead of instruments of war, humanity used the rockets and computers of World War II to orbit the Earth and land on the Moon. Humanity had taken its ingenuity and moved in a better direction.

Because of this, twenty-four years after lying full of hate in a hospital bed, Elly saw humanity walk on the Moon. As he did so, he was sitting in a beautiful home filled with pictures of his loved ones, a picture-perfect life built through love. For Elly, the moon landing will always represent what happens when you turn away from hate. He knows better than anyone that only when you give up hate can you truly start to live.

CHAPTER 2

Yenny, Chile

Imagine finding yourself in a completely uninviting environment of rigid and coarse surfaces. A landscape with no presence of life, no signs of water, a terrain completely devoid of all movement. No clouds to block the sun's rays, and a harsh temperature difference between day and night... if you make it that long. A powder-like dust covers most of the ground, creeping up the higher ground and filling in the depressions. Bold mountain peaks are visible in the distance while rounded hills stand in the midground, as if providing a warning to turn back. Trying to look for a horizon is impossible, the distance interrupted by rising cliffs in all directions, blocking sight of what may lie beyond.

"It really does! It looks just like Valle de la Luna!" Yenny's mother said, sitting on one of the little old wooden chairs beside the couch. The family was gathered at the TV. As the only kid in the house, ten-year-old Yenny had the prime spot, seated on the big living room sofa between her grandmother and her uncle. Her legs were dangling down, swinging excitedly a few inches short of the red concrete floor as she listened to the adults compare the alien landscape to their local desert. Indeed, much of Chile's landscape, running along the side of South America's west coast, resembles some of the vast, open terrain of our solar system.

Chile is often described as being wedged between "the deepest ocean and the tallest mountains." Satellite imagery reveals a drastic shift from the lush Amazon rainforest at the center of the continent to the harsh, arid desert on the other side of the north-south running Andes Mountains. Chile's narrow territory rests entirely west of this gigantic mountain range, largely separated from the lively jungles of the rest of the continent.

The Valle de la Luna is a rocky portion of the Atacama Desert, the driest non-polar desert in the world. Its unwelcoming environment makes up a significant portion of Chile's northern half. For planetary scientists, Valle de la Luna is well known for its role as an analogue site—sites on Earth used to replicate conditions on other worlds, providing the closest conditions we can find here to other places in our solar system. In this "Valley of the Moon," researchers from space agencies and universities all around the world test robotic rovers and train astronauts in preparation for the exploration of the moon or Mars.

As Yenny's family watched the scenes unfold from the alien world, the eldest woman of the house, Yenny's grandmother, teared up with joy—though perhaps trying to cover this up out of slight embarrassment. "See, you lived to see a man on the Moon, we told you!" her son, Yenny's uncle, teased. In the days leading up to the landing, and during the dinner immediately before, Yenny's grandmother had expressed disbelief that people would actually be landing on the Moon. To her, it seemed impossible. Sure, people can fly around in rockets, but no one was going to actually go all the way to the Moon! As the astronauts filmed their leaping steps and panned the camera across the lunar desert, Yenny's grandmother's doubts quickly dissolved.

Yenny herself had sat next to her grandmother quietly up until now. Inside, she was not only excited to see that people were on the Moon, but was also jealous of how fun it looked. Finally, she blurted out: "I want to do that! I want to stand on it and bounce around!" Between the family members, there were many similarly excited statements exchanged. Of course, all the while their eyes were fixed on the TV, listening to each other, but not looking away from the old RCA set beaming grainy black-and-white footage from the Moon. The seemingly impossible feat was being received by the TV's one-and-a-half-foot-long rabbit-ear antennas pointed in the air; the magic of that was lost on no one.

Ideas and questions rushed through her mind, but tonight Yenny knew she didn't have an undivided audience for her thoughts. That was fair; the whole family was mesmerized by what was happening. In a few weeks however, Yenny would have an audience: She, like every other kid in her fifth grade class, was going to need to present a project that expressed how she felt when she saw a man on the Moon.

•••

Despite the fact that she lived with her grandmother and was, by her own admission, "a bit of a nerd," Yenny's lifestyle in Viña del Mar was pretty typical for Chilean kids her age in the late 1960s. An open-concept ground floor greeted her as she came downstairs each morning for breakfast. It was centered around the kitchen table, the nexus of socialization, where family members, neighbors, and friends would gather.

Concrete made up the walls and floor of the ground level. Upstairs, a wooden addition housed the bedrooms. Because Viña del Mar, like all of Chile, was prone to earthquakes, most houses were built in this configuration. The concrete bottom floor needed to be able to withstand the violent shaking of the ground that can accompany movements between the Nazca and South American tectonic plates. The motion of these two plates is what gives rise to the massive and beautiful Andes mountain range, but also causes frequent, strong earthquakes. The largest earthquake in recorded history was a magnitude 9.5 quake in Chile in 1960, but the country also experiences hundreds of smaller quakes every year.

After breakfast, Yenny would go to school for the first session of the day. Her school had classes until noon, when the students returned home to have lunch with their families. After lunch it was right back to school until about 3:00 p.m. Because she was living with her grandmother and there were no other kids in the house, Yenny's after school routine was perhaps where things were a bit different from her classmates. There was none of the loud chaos that can come with a house full of kids, no siblings to fight with. Instead she would spend time doing homework and practicing her cursive writing at her grandmother's insistence. When Yenny's hand got tired she would take a break and look at a globe on her bedroom desk, running her fingers over the raised mountains down the South American continent.

When her practice sentences were complete, she often left the house to see which kids were outside to play with. Surely, as other kids finished their homework or chores, they filtered outside too and spent a few hours riding bikes together and playing in nearby parks. As long as everyone was home in time for their dinner, everything was okay. In Chile it was common for families to eat later, around nine or ten o'clock, giving kids more time to play outside together.

While many kids around the world—and certainly in the United States—were talking about rockets and astronauts during their bike rides and park games in the sixties, in Viña del Mar this was not really the case for Yenny. The Space Age was underway, but in the small suburb the advancements of space travel in America and the Soviet Union were not regular news. Even in school it wasn't talked about much.

That was, until the week before winter break, 1969.

In June, before the students were to go away for a few weeks, it began to circulate around school and around town: there was going to be a man on the Moon. The teachers told the kids to pay attention, watch or listen to the event, and be ready to provide a report on it when classes resumed after the break.

Because the lead-up of increasingly difficult missions was not as prominent in Chilean news at the time, the announcement of a man on the Moon seemed to come out of nowhere. Influenced by her grandmother's disbelief at what she was hearing, Yenny too had her doubts. How could they do that, go all the way to the Moon and show it on TV? But everyone Yenny knew—family members, teachers, and neighbors—was talking about it. With some of them, like her grandmother, doubting it would happen, Yenny didn't know what to believe. She would have to watch it herself to find out if it was real.

•••

As the day approached, signs emerged that the rumors must be true. Now that it was winter break, Yenny was free to spend a lot more time with her friends and neighbors, often listening to music inside when the weather wasn't so great. Along the way to one of their houses on July 16, Yenny and her friends biked past a group of people standing outside an electronics store. They were watching the TVs in the window. The news was rebroadcasting the launch of Apollo 11, which had happened earlier that morning.

This, as clichéd as it now seems, was a reality for much of the 1950s and 1960s all around the world. In the case of Chile in 1969, it was not common for many people to have televisions in their homes, and even if they did, the TV stations didn't broadcast until the afternoon. Lots of people wouldn't see anything on TV until they were out of the house and out for lunch. For people who didn't own a TV, it was only in restaurants, stores, or bars that they could see these broadcasts.

It's easy to imagine that, whether you owned a TV or not, it would surely catch your attention to walk past a window and see the Saturn V—a building-sized rocket—soaring through the air. In Viña del Mar, this type of unifying attention was usually reserved for football games, but Apollo was special. All around the world the launch of Apollo 11 was attracting audiences, and nowhere more so than in central Florida, thousands of kilometers to the north of Yenny and her friends.

People began to gather a day earlier in the coastal islands, peninsulas, and beaches surrounding Cape Kennedy, where NASA would launch its Saturn V rocket to the moon. At the time, it was estimated that up to a million people came to watch the event in person. The flat, sandy landscape of the sunny Florida coast made the viewing area tens of kilometers in size. Cape Kennedy sits along the east coast of the state, about halfway along its long Atlantic coastline. The US had been testing rockets in the area since the 1950s, but when the Apollo program ramped up in the mid-1960s, the entire region turned into a haven for space activity, becoming the aptly named "Space Coast" that it remains today.

The cape itself extends outwards along the coastline from the north, pulling away from the mainland as it heads towards the ocean. The gap between the cape and the mainland coast includes Merritt Island as well as a host of smaller islands and protruding sandy points. It was along all these slivers of land that hundreds of thousands of people gathered to see the biggest rocket ever built soar above the Atlantic Ocean on its way to deliver humanity's first lunar tourists.

With Florida's wetlands in the background, the Saturn V launches from Cape Kennedy with over a million people watching from the surrounding area.

+ For All Humankind +

Four days later, it was evening again and Yenny was just getting home from a bike ride. She and her friends had been talking all day about what they expected to happen on TV that night. She hopped off her bike and left it on the grass in the front yard, running up to the front door. Adding to the excitement was the fact that her mother and uncle were inside when she came in. They had come over from the nearby town of Valparaiso for dinner and to watch what would unfold. Tonight, the topic of discussion during dinner was singular. The conversation lively as usual, but perhaps a bit more hurried than normal. Yenny sat as she usually did, eating politely and listening to the grown-ups.

"How are they getting all the way to the Moon?"

"Are they going to bring pieces back?"

"What will it look like?"

When everyone was done eating, they rushed through the clean up and went to the living room.

An earthquake a few days earlier had moved the position of the television's antenna, so Yenny's mother adjusted the metal rabbit-ears slightly to improve the image of the newscast. It was nighttime now and the house was mostly dark, save the living room section of the main floor. Guitars hanging on display, and for easy access for her uncle's playing, lined the back wall of the living room behind the couch. Resting between two giant plants, the TV was the focus of attention in front. The screen's glow bounced off of the nearby aquarium and reflected onto the window on the opposite side in a curvy, undulating pattern picked up off the water.

Yenny sat with her grandmother on the couch. On the other side, her uncle was still trying to convince the elder of the family that it would actually happen—a person would walk on the Moon. For her part, Yenny wondered: what would she and the other kids talk about when they went back to school if no one ended up walking on the Moon?

•••

After the landing, excitement aside, Yenny could tell that her mom and uncle were ready to go back home. They would have to work the next day and it was now near one o'clock in the morning. After a hug and kiss goodbye for each departing guest, Yenny and her grandmother went upstairs for the night. As she got tucked in, Yenny tried to calm down. They had really done it! People had gone to the Moon, and her whole family got to watch on TV as it happened!

Over the next week, it felt like *everyone* was talking about the men on the Moon. But as much as she loved talking about her favorite moments from the Moonwalk, Yenny was starting to get nervous about the presentation. She wanted to do very well, but was unsure exactly what she would talk about. How could she fully express in writing the excitement she felt that night?

She put her family to work, asking her mother, uncles, aunties, and grandmother to collect as many Moon stories as they could from the newspapers. When family members were done reading the local papers, *El Mercurio* and *La Tercera*, they would deliver them to the house where Yenny would go through the articles and cut out any and all pictures having anything to do with the Moon.

With a pile of pictures, an idea dawned on Yenny and her mother: What if they presented their project as a letter to Earth from the Moon, as if to say, "Here are the pictures from when you visited me, come back soon!"

It was perfect. She and her mom went out and got a large piece of bright-blue Bristol board and folded it into an envelope. They colored the outside with a ribbon and addressed it:

To: Earth
From: The Moon

Inside were all the newspaper pictures from humanity's first trip to the Moon.

When school resumed after the winter break, Yenny carried her bright blue Moon letter to class. On the first day back, she presented to her classmates and the parents in attendance. "When the moon landing was on television, it felt like the Moon was sending us a letter, a sort of 'hello'," she said. She also relayed that when she and her family saw the moon on TV, they immediately thought of Chile, saying "yes, it does look like the Valle de la Luna!"

Finally, when asked by her teacher if she would like to go to the Moon, Yenny remembered the bouncing. "Yes, I want to bounce on the Moon like the men did!"

Though the younger generation were excited to think about when they might get to bounce around on the Moon, there were always those who weren't so impressed. Some of people from Yenny's school, church, family, and neighborhood thought that the moon, and space, was not somewhere people should go; that it was not right for humans to venture to the heavens. Many of these people were older, like Yenny's grandmother. But while her grandmother was vocal in her disbelief that it would happen, she herself was not so concerned. Once she was happily proven wrong that humans couldn't truly land on the Moon, she was glad to see it. She had the same mentality as the kids: what's next?

•••

In the fifty years since Apollo 11's launch, humanity's scientific and technological advancements have been nothing short of astounding. In the field of space exploration alone, we have landed robots on Mars, flown past the furthest bodies of the solar system, detected planets orbiting distant stars, and had astronauts living in orbit around Earth continuously for two decades onboard the International Space Station (ISS). This is to say nothing of the countless advancements in medicine, communication, and transportation achieved in the same period. At the same time, hundreds of brilliant symphonies have been written, countless masterpieces painted, and libraries upon libraries of beautiful books written. Humanity keeps pushing ahead, aiming to accomplish new things.

It is these advancements that Yenny thinks about today when asked about Apollo 11. And while some may feel that five decades is a long time, to Yenny, when you live it, it doesn't feel that way at all. "It's not a very long time if you think about it," she says. And it certainly isn't if you compare it to other fifty-year periods of the past. In previous centuries, fifty years of science, or art, may have seen limited advancements, not a long enough period for major leaps in human capacity. But not in the twentieth and twenty-first centuries, a period in which human ingenuity has gone into overdrive. That is what Apollo represents to

Yenny—a harbinger of the rapid advancement that humanity now undertakes.

Of course, Project Apollo didn't cause *all* of our advancements. It was just one piece of the pie. But it was certainly a marker of what was to come. It ushered out the old and brought in the new. In the following decade, the microchip would be invented, providing the potential for miniaturized computers that could fit in people's homes (in the 1960s computers were, on average, the size of a small bus). The 1970s, too, saw the earliest versions of the millennium-defining invention of them all: the internet.

The rapid advancements of Project Apollo and in the post-Apollo era also help to explain some of the difficulty many people, including Yenny's grandmother, had in believing that a Moon landing was even possible. Humans had only gotten off the ground with flight in the early twentieth century. For an older generation at the time of Apollo, it's understandable that trips all the way to the moon seemed impossible in the same lifetime.

This is a good reminder not to expect the future to always look like the past. Advancements can be difficult, if not impossible, to predict. It's always best to keep an open mind as to what is possible. Yenny wants kids today to know this. Even if something seems impossible, it could still happen in your lifetime or your children's—the cure for terrible diseases, the end of environmental problems, equality for all and world peace—it can all happen. We should never give up and think them impossible.

Yenny's own story reminds us that these rapid advancements don't only occur on the scale of human history as a whole. They include people's individual lives and family histories too. Yenny's grandfather traveled from Palestine to Chile by boat in the 1920s. The trip took three months, during which time he learned three languages along the way. He accomplished this without any of the technological assistance we enjoy when learning languages today. What's truly amazing is that, with hindsight, we can say that many of the passengers on that slow-moving, oil-powered boat would live to have a story about the time people walked on the Moon.

By the time Yenny recounted her own Apollo 11 story, she sat with her daughter Kristen in Canada. Though born two decades after the landing, Kristen knows all about NASA's Moon program. She trained at the very same NASA facilities in Houston, Texas as the Apollo astronauts. Today, she is a Flight Controller with the Canadian Space Agency, and works to help astronauts from around the world prepare for their time on the ISS—especially if they have to operate the Canadian-built robotic arm aboard the station.

Yenny reminds us: "Appreciate what you have and believe anything is possible." We cannot give up on goals just because today we think they are impossible. The things we consider possible today will help define what the next generation will be able to achieve tomorrow. In the span of four generations, one family went from traveling across the ocean in three months to operating robotics on a space station that orbits the entire planet every ninety minutes.

One small voyage for a man, one giant leap for the whole family.

CHAPTER 3

Abdel, Sudan

Moonlight provided the only illumination in the darkness of rural Sudanese nights. On this particular night, the rays reflecting off the puddles of water from the day's rainfall provided an extra glow. The village in which young Abdel lived had no electricity. No streetlights shone on the dirt pathways. The occasional candle might be seen in the window of a house here and there, but once the hour was late enough, even the candles were extinguished.

Amidst this near-darkness, when there was a Full Moon, the children would come out to play.

On this night, kids were splashing around in the puddles of rain, laughing and squealing jubilantly without an adult in sight. Exploration and play were encouraged here, fostered by a trust that the children would be all right; they were under the helpful eye of the Moon. But Abdel would often get distracted from the games being played. He would gaze up at the Moon and see a giant question mark. To him, there was a mystery hanging in the sky.

In Islamic culture, the Moon is honored and respected. It is seen as a symbol of serenity and beauty, a source of inspiration. Love poems across the centuries effuse romanticism for women described as being "as beautiful as the Moon." Poetry of the famous thirteenth-century Persian Muslim poet Rumi often refers to the Moon to evoke beauty and serenity.

Navigation also relied heavily on the Moon and stars. With little in the way of visual markers on the ground in the expansive desert, the Moon, Venus, and the North Star (Polaris) were crucial for caravans riding camels, traveling from village to village.

The Moon also provides the basis for the Islamic calendar. This calendar, called the Hijri, is based on lunar months marked by the Moon's phases. When the first crescent begins to appear after the New Moon, a new lunar month begins. The Moon is at its brightest during the twelfth through the seventeenth day of each lunar month, with the Full Moon marking the fourteenth day.

The Full Moon always seemed to bring some sort of adventure to Abdel's life. On another bright night in his childhood, Abdel and his friends were outside caring for his family's sheep. Though they were young, they were entrusted with keeping the flock safe from predators. Typically, these nights were uneventful, save the antics of little boys. However, the bright Moon provided benefits for more than just the humans of the area. This night, their conversation was abruptly stopped by one member of the group.

"Hey! Look behind you!"

The young men turned away from the flock to see a large wildcat stalking the sheep from the grass, only a few meters from where they were all standing. The young men were face to face with one of Earth's most dangerous natural hunters. Regardless of whether it was a cheetah, leopard, or, most terrifyingly, a lioness, these four humans would have been no match for any of the big cats that roam Sudan.

But no fear set into them. Instead, their duty to protect the sheep from being eaten was at the forefront of their minds. They collectively began shouting at the animal at the top of their lungs, showing no fear toward the carnivorous cat. It was a tactic they had all learned from their elders. In Sudan, your best chance against large predators was to bind together and get loud. Luckily for them—and the sheep—the noises were indeed enough to spook the animal and deter it from its late-night snacking plans.

It must not have been too hungry.

•••

Abdel was in quite a lucky position to indulge his curiosities in life. Growing up in the 1950s around the time Sudan was gaining its independence from Britain, an education was not easy to come by. Schools were limited, and getting a seat in the local classroom was based on a lottery system. If your number was drawn out of a bowl, you got the prize of an education. If your number wasn't selected, you were sent home. Quite literally, Abdel had his lucky number picked; the year he began school, only forty or so students were selected among hundreds hoping to receive an education.

Thanks to being selected, Abdel learned to become an avid reader. A mobile library came to his secondary school every two weeks, brandishing new doses of knowledge for his mind to ingest. He also grew up in an environment without boundaries, where he was allowed to go wherever he wanted to go, do whatever he wanted to do. This allowed him to explore and forced him to learn how to solve problems. He loved to tinker, to figure out how things worked around him. The opportunity to explore, unguided, along with access to a library, primed Abdel to become one thing: a professional problem solver—also known as an engineer.

•••

The University of Khartoum was initially founded by the British in 1902 as Gordon Memorial College. As it grew, it quickly became a hub for studies such as biology, medicine, and engineering. By July 20, 1969, Abdel was a twenty-two-year-old engineering student at the university. On the Hijri calendar, it was the first of Jumada Al-Awwal—the fifth month of the year 1389H. It seemed quite fitting to him that the first day of a new lunar month would be the day a crew of three humans would launch aboard the mighty

Saturn V rocket and attempt humanity's first landing on another world.

All international news came to Sudan via the BBC, and predominantly via radio. Televisions weren't yet commonplace; at the university, there was only one on the entire campus. This television was in the student union, a common place for students to gather and socialize. At the time, the "weekend" in Sudan only consisted of one day: Al-Jumu'ah, the Hijri calendar equivalent of Friday. On these nights, Abdel and the other students would typically gather in the student union in the evening of Al-Khamīs (Thursday) to watch the television.

But this week was different. It was late on Al-Ahad (Sunday), and students were clustered around the single black-and-white television. Abdel had made sure to seek out a prime viewing spot so as not to miss a single bit of information the BBC newscasters were conveying. People were mostly quiet. Everyone knew they were experiencing an important landmark for humanity. For Abdel, it felt appropriate to take it in together, gathered as students in this communal space, to see the mission carried out.

At 10:17 p.m. in Khartoum, Sudan, Neil Armstrong's words echoed from the Moon across the Earth: "Houston, Tranquility Base here. The *Eagle* has landed."

Charlie Duke, the Capsule Commander on the ground in Houston, replied with a statement that likely represented the feelings of almost everyone that had tuned into the landing as it happened: "Roger, Tranquility Base, we copy you on the ground. You've got a bunch of guys about to turn blue. We're breathing again. Thanks a lot."

The nervousness had been inspired by a series of alarms going off in the Lunar Module during Neil and Buzz's lunar descent. As the astronauts were only tens of meters above the surface, the computer flashed a series of 1202 alarms, indicating that the computer was ignoring less important commands in favor of critical ones. Old computers did this to let you know you were risking overloading them. However, upon first hearing Neil Armstrong report "we have a 1202 alarm," ground control at Houston didn't know which exact problem the computer was signalling.

The Lunar Module's onboard computer could relay hundreds of different alarms. It took a team of computer engineers to know them, and even they had to rapidly look them up in massive binders. However, given the timing of the alarm, there was practically no time. Neil and Buzz needed to know within a few seconds. If it were a serious alarm, they needed to abort as fast as humanly possible. Had they been told too late that something was wrong with the spacecraft, they might not have enough fuel to get back up to the Command

Module. Unable to land either, they would have been stuck between a rock and empty space.

Luckily, this exact alarm had come up in the last week of flight simulations, and a young computer engineer quickly relayed that the alarm could be switched off. It was a sort of courtesy alarm, but was not an indication of anything mission threatening. After all this, it's easy to understand Duke's relieved message back to Armstrong upon landing.

384,000 kilometers from the site of the problem, Apollo flight control staff had to act quick to save the day.

It would be another six hours and thirty-nine minutes before Neil Armstrong and Buzz Aldrin exited the Lunar Module and set foot on the Moon's desolate, gray surface.

After the landing had been successfully confirmed, Abdel ran all the way from the university back to his uncle's home, fueled entirely by adrenaline over what he'd just witnessed. By now it was dark, well past evening twilight. Bursting through the front door of the house, he found his uncle in the midst of his *Isha'a* evening prayer, supplicating on his prayer mat.

"Uncle! Uncle! There's some great news today! Men landed on the Moon!" Abdel exclaimed, unable to contain himself until the prayers were complete.

His uncle looked up at him with a mixture of annoyance at his interruption and disbelief at his statement. "What?!" he shouted. "What do you mean?"

"The Americans!" Abdel explained breathlessly, trying to muster the breath to speak. "They had a mission, they sent men to the Moon!"

Still not buying it, his uncle replied, "They are liars!" To him, the idea of humans being able to reach the Moon was utterly preposterous. Surely his nephew was attempting a joke, or perhaps confusing something theoretical he'd discussed in class with reality.

Realizing he wouldn't be able to convince his uncle on his own, Abdel walked to the household radio and turned on the BBC. The exasperation on his uncle's face slowly transformed into shock as he listened to the newscasters recap the events of the landing. But he still didn't quite believe what he was hearing. "Why don't they tie a rope to it and bring it down to Earth where we can see it for ourselves?" his uncle said, somewhat sarcastically.

Abdel was disappointed that his uncle wasn't fully appreciative of the enormity of what had happened that day. His own enthusiasm did not wane however, and he continued to listen to BBC Radio for hours, waiting for Armstrong and Aldrin to take their first steps on the lunar surface.

Many others across Sudan were equally skeptical, but as NASA continued to send humans to the Moon on subsequent Apollo missions over the next few years, more and more of the general population began to recognize what was actually happening. Humans *really were* walking on the Moon!

Apollo fever gripped the world, and Sudan was no exception. Shops and businesses with names like "Apollo Café" and "Apollo Bus Lines" popped up in Khartoum. People buzzed with speculation and wonder about what was going to happen next in this Space Age.

Eventually, Sudan followed its own space dreams. In 1977, the government created the National Remote Sensing Center, later renamed the Remote Sensing and Seismology Authority (RSSA). Perhaps inspired by the views of Earth captured by the Apollo missions showing our fragile planet isolated in the darkness of space, the RSSA was formed with a goal of—among other things—using satellite data to protect the environment and monitor natural disasters. Much more recently, Sudan's National Research Centre established the Institute of Space Research and Aerospace to promote space science studies and technology development.

Abdel's alma mater, the University of Khartoum, has continued to be an engineering powerhouse in the country and around the world, making its way into the space sector as well. The university's Space Research Centre has been developing CubeSat prototypes—small satellites that take advantage of the tiny technology created for things like cell phones—as well as satellite tracking ground stations with which to eventually talk to the CubeSats after launch.

CubeSats have heralded a revolution in space exploration and Earth observation due to their small size and use of many components that can be readily purchased off the shelf at stores. While traditional satellites are often the size of minivans or even school buses and require custom made parts, CubeSats can be as small as a coffee mug or a loaf of bread and weight just a few pounds. This makes them quick to build and inexpensive (in the relative sense) to launch, opening up the possibility for more people and countries around the world to be able to build their own satellites. What was previously an expensive endeavor only attainable by the largest governments in the world, is now becoming accessible to innovative students in college and even high school! The benefit: faster, clearer images of our changing Earth, and knowledge needed to keep our environment habitable.

•••

Fifty years have passed, and Abdel is now a professor of mechanical engineering in the US. The human achievements of Apollo, the advances it spurred on in science and technology, and its legacy of inspiration are the things he passes along to his students.

Apollo proved to us that humans can conquer anything if they put their minds to it. "Humans…we're tough nuts to crack. But we need a vision and the means to accomplish it," Abdel states passionately. He cites President Kennedy's speech challenging us to go to the Moon as a prime example of this. "Before Apollo, I did not realize the potential humanity has to attack a very challenging problem and solve it."

Smiling in his university office, Abdel dreams of living long enough to see humans set foot on Mars. As a child in Sudan, without electricity, before humans had ever been into space, he could not have imagined the progress that happened so rapidly over the course of his life. In his classroom, he tries to convey this to his students, telling them that the progress and problems of the future may be unimaginable today.

Abdel's view of the future is a positive one. "The future is bright. The only problem is that we need to figure out how to be good to each other, to make sure we make positive impacts on one another and minimize the negative impacts."

To Abdel, education is the most important tool in solving this world's problems. He worries that the education system today boxes students in to be career-focused, rather than encouraging them to explore, to solve problems, to collaborate together. "Education is the catalyst for dreams," Abdel says emphatically. "We should not leave *anybody* behind without helping them get educated."

Having grown up in a time and place where he was lucky to have received an education at all—literally the random luck of the draw—the opportunity which access to an education provides people with is all too clear to Abdel. Ultimately, we need to eliminate the role of luck in people's opportunities to learn. Luck will inevitably play a role in our lives, but we need to assure everyone has the knowledge necessary to take the utmost advantage of good luck when it comes. Whether it is scaring aware a predator, becoming an engineer, or landing on the Moon, with a little bit of luck, knowledge can take us anywhere.

CHAPTER 4

Phil, England

Local mythos says the location of the southern English city of Salisbury was determined by an archer who randomly shot an arrow from the top of a hill down into the adjacent valley. Like most myths, this probably isn't what actually happened, but it makes for a fun origin story.

Salisbury doesn't really need to rely on mythology to be interesting though. It has plenty of true things about it that make for stories just as good. At the center of town stands Salisbury Cathedral, supposedly constructed at the spot in the valley where the archer's arrow landed. Built over eight hundred years ago, it is among a number of other similarly old, similarly inspiring, architectural masterpieces in the city square.

The city is probably just as famous for what's inside of it as for what's outside. Pastoral rolling hills fitting the stereotype of the quaint English countryside surround the city. Just beyond the city limits is one of England's most attractive tourist sites: the enigmatic Stonehenge. The site is famous for a series of mysterious stone slabs set up in a circle about five thousand years ago. No one is sure exactly why it's there, or how it was even built. Some of the stones weigh as much as five tons, and were transported all the way from Wales—over 250 kilometers away! How humans five thousand years ago managed to move such large stones over such a vast distance remains a bit of a mystery to this day.

Thanks to its distance from the coast, Salisbury was lucky to never be a target of Nazi attacks during World War II. Sadly though, this doesn't mean the war didn't affect the city. As V2 rockets bombarded London—England's capital city, about 150 kilometers northeast of Salisbury—thousands of Londoners fled to small towns like Salisbury to avoid "the Blitz." At the same time, thousands of brave young people from the English countryside, including Salisbury, made their way to the European and African fronts to serve in the effort against Hitler's army.

After the war, Salisbury and the rest of England went through significant changes in an era of rebuilding. As the world entered the second half of the twentieth century, things were evolving rapidly. Humanity had both split the atom and sent machines into space. Throughout the late 1950s and early 1960s, newspapers were filled with increasingly incredible stories about skyscraper-sized rockets, nuclear-powered machines, and chimpanzees being launched into space. Living in this evolving postwar world, a young Phil Stooke was fascinated by the stories he saw in newspapers and on TV. What ten-year-old *wouldn't* be interested in space-traveling chimps?

But it was during one particular Christmas break that Phil really began paying attention to space. In December 1968, humans traveled to the Moon for the first time on NASA's Apollo 8 mission. The three astronauts onboard didn't *land* on the Moon on this particular mission. Instead, they flew around it to make sure the Apollo spacecraft worked in the tough conditions of lunar space. Their main goal was to ensure they could make the journey all the way to the Moon and back without having the spacecraft lose power, life support, or communications—pretty important stuff. Before Apollo 8, every other crewed space mission had flown just above the Earth's atmosphere. This was the first time humans ventured beyond Earth orbit. When the astronauts flew behind the Moon, they could no longer even *see* Earth. These astronauts were the first humans *ever* to experience a moment with no sensory connection to Earth.

During that formative Christmas break, Phil saw news about the Apollo 8 mission every night. He was hooked. While interests tend to come and go when you're a kid, this was no passing fad. As he got older, anything and everything space-related started catching his attention. He read science fiction books by the likes of Arthur C. Clarke and Robert Heinlein—books describing different planets, real and imagined, in painstaking detail.

In hindsight, their descriptions of planets like Venus and Mars turned out to be laughably inaccurate, but no one knew that in the early 1960s. For kids like Phil, these books might have been predicting the future, for all they knew. And with each newspaper article about the American and Soviet space programs, it felt like the space travel described in science fiction books was getting closer to real life. Science fiction and reality appeared to be on a collision course.

Three months after Apollo 8, the Apollo 9 mission launched. This mission didn't go back to the Moon—it only flew around Earth. The goal of Apollo 9 was for the crew to practice a series of crucial moves that were too dangerous to test for the first time all the way out at the Moon. This meant docking the Lunar Module with the Command/Service Module, separating, reconnecting again, and separating the Command/Service Module. Back in Salisbury, Phil was also learning these maneuvers himself from articles and diagrams in magazines like *Colliers*, *Sky & Telescope*, and *Popular Mechanics*. His fascination and expertise growing with each article and each Apollo mission, his stack of space magazines started to reach the height of his bed.

In May of 1969, Apollo 10 was ready to go. This mission would be the so-called dress rehearsal for the actual Moon landing. It combined the activities of Apollos 8 and 9 with a first test of the Lunar Module at the Moon. Not only was NASA advancing its expertise—so was Phil; by the time he watched the Apollo 10 launch, he was a full-fledged teenage Moon expert, with magazines now stacked to the height of the old wooden desk in his bedroom.

But by now, magazines weren't enough.

Hungry for all things space, Phil's fascination drove him to do what a lot of people did at the time: he started cutting out space pictures and articles from newspapers to keep in a scrapbook. If you were born after the advent of the internet, scrapbooks like this might seem like a strange idea. Today we can see any picture we want, at any time, on our phone or tablet. But there was a time when things like pictures of astronauts, your favorite band, or your favorite athlete were actually hard to find or expensive to get. Unless you owned books or saved newspaper and magazine articles about these things, you couldn't read about them whenever you wanted. So, when you saw an article about something you liked, it was totally normal to cut it out and paste it into a scrapbook.

However, Phil didn't have to rely solely on his scrapbooks to follow the Apollo program. After a few years of space fascination, he got lucky enough to score access to a book full of photographs directly from the Moon. While in secondary school, a friend was able to borrow a Moon atlas from the military archives where his father worked and pass it along to Phil. In the atlas were close-up images of the Moon taken by the American Ranger 7 spacecraft.

Launched in 1964, Ranger 7 was designed to *intentionally* crash into the Moon. While that may seem like a waste of a Moon probe, it was really all NASA could do in the early 1960s. Expertise in remote controlling a space probe was far from where it is today. At the time, the only way to get close-up pictures of the Moon was to crash into it while taking pictures on the way down—a one-way trip for the sake of science.

Some early lunar missions actually missed their (rather large) target entirely and flew right past the Moon. Ranger 7 on the other hand was successful in smashing into the lunar surface four times faster than a speeding bullet. The close-up photos Ranger 7 sent back right before impact were amazing, but before now Phil had only seen one or two of them that were shown on TV or printed in newspapers. The moon atlas, on the other hand, had *all* of them. From these pictures, Phil learned that there are craters almost everywhere on the Moon. This, of course, meant that landing there was going to be tricky.

Places with lots of craters are not somewhere you necessarily want to try to land a spacecraft for the first time. Aside from the risk of falling into the craters themselves, the terrain around craters tends to be uneven, hilly, and strewn with giant boulders shot out by the impact that formed the crater in the first place. One of the first challenges for NASA's Moon landing desires was finding a safe place for humans to land. This objective was the main reason for sending missions like Ranger 7.

Between classes, Phil pored over and thoroughly studied the images from the atlas. He started wondering: Why are some areas smooth and others much rougher? Would it be safer to land inside a giant crater, or in an area with a lot of little craters that aren't as deep? As he looked at the images of the moon's alien surface, teenage Phil was *actually doing* space exploration. And at the time, he didn't realize that scientists in the Soviet Union and America were doing exactly what he was. NASA mission planners were even relying on a lot of the same photos that were in that atlas.

Even though Phil shared his pastime with experts at NASA, he didn't feel much like he was a part of the Space Age. As far as he could tell, there weren't many other kids in Salisbury that were all that interested in space exploration. The world of cosmonauts, NASA, and the Moon felt a galaxy away. Space exploration seemed only practical as a hobby to him. He watched from afar and collected news articles, magazines and books. Really, it seemed to Phil that he felt about pictures of the Moon the same way his classmates did about pictures of The Beatles, or the movie stars who played James Bond and Dr. Who. The idea of meeting or working with these people he saw in magazines, whether they were astronauts, rocket engineers, or planetary scientists, seemed an unimaginable path.

After the school semester, and reluctantly returning the Ranger 7 atlas to his classmate's father, seventeen-year-old Phil was on summer break from secondary school. He spent no small part of his vacation going through his scrapbooks, reading about the Apollo missions, and following the nightly news for updates. The first Moon landing was only days away. Surprisingly, there didn't seem to be many other people in town as excited about the Apollo 11 mission as he was. There was some chatter about it for sure, but many people seemed to be paying no attention at all. Obviously to a teenage Apollo expert with a veritable library of moon scrapbooks and magazines, this was unbelievable—in a few days, there would be people walking on the Moon for the first time in human history! How was the enormity of this moment not resonating with more of his friends and family?!

On July 20, 1969, Phil had to wait most of the day before anything about the Moon landing came on TV. The mission had been strategically timed so that the Moonwalk (fancily known by NASA as an Extravehicular Activity, or "EVA") would take place during prime time on the US East Coast. For Phil, who was five hours ahead of that time zone, this meant that the landing would happen around 9:00 p.m., and the Moonwalk sometime after midnight. Luckily, he was on summer break and there was no need to worry about things like sleep.

To take it all in, Phil made his way into the living room on the main floor and turned on the TV, changing the channel to the BBC. He pulled his favourite chair a bit closer to the screen than usual, taking his front row seat at one of the most important moments of the century. By this time Neil, Buzz, and Mike were already in lunar orbit, getting ready to part ways by splitting the Moon-landing Lunar Module (LEM) from the Moon-orbiting Command Module.

Phil watched as the newscasters prepared viewers for the eventual landing and Moonwalk, explaining what would happen, and at roughly what time. The TV schedule ultimately depended on the comfort of the astronauts and decisions at Apollo mission control. In England, people saw a panel of experts sitting in the BBC studio, kind of like the intermission of a World Cup game. The evening was hosted by Patrick Moore, a famous English science journalist. Much of the coverage that night was about the three astronauts. Like the rest of the Apollo crews before this one, Phil already knew these astronauts' names and little bits of their history and personalities.

One thing that was the same in the TV coverage of the landing around the world was the use of models and cartoons to show what was happening during the mission. There were no cameras filming much of the spacecraft maneuvers, and computer animation didn't exist yet. Instead, news anchors would demonstrate the maneuvers of

✦ For All Humankind ✦

the lunar lander and command module with models, like kids playing with toys. There were even a few TV stations that built life-size replicas of the twenty-two-foot (6.7-meter) tall lunar lander.

As the astronauts completed inspecting the LEM (nicknamed *Eagle*) and preparing for undocking, the audio from the astronauts and NASA's ground control in Houston began to play over the news broadcast. Around 6:45 p.m. in Salisbury, the two ships—the *Columbia* command module and the *Eagle* lander—separated from each other. For Phil, this is where the adventure really began. He knew that almost everything from this moment onward was being done for the first time. His eyes would not stray far from the TV screen until this next part was over.

It would take a little over an hour for the lander with Neil and Buzz in it to descend all the way to the Moon's surface. Phil began to get a little nervous. The lander had never flown so far away from its partner spacecraft, the Command Module *Columbia*.

Even scarier, the lander had never landed on the Moon. In fact, it had barely landed on Earth.

The *Eagle* was designed for one purpose: to land on the Moon. It wasn't aerodynamically designed to fly in Earth's atmosphere, or in Earth gravity. The moon has one-sixth the gravity of Earth, and no atmosphere. There was no way for NASA to perfectly recreate these conditions on Earth to fully test the lunar lander. So, for the astronauts, learning to fly the *Eagle* was like learning to drive a boat without ever being on the water.

While NASA engineers did figure out a way to test the *Eagle* lander in conditions *close* to that of the Moon, these test flights didn't always go well. Just one year before Apollo 11, a landing test vehicle flown on Earth had a serious malfunction and crashed. The pilot had to eject to save his life. That pilot was Neil Armstrong—the same man now trying to land the same spacecraft on the moon.

From an outside perspective, the scene was of a seventeen-year-old staring intently at an old 1960s TV in a quaint, peaceful English living room. But inside Phil's head, tension was high. If Apollo 11 were to experience catastrophic failure, it would likely be during this part of the mission.

Some signs of trouble did arise. As the lander descended, the astronauts heard a series of alarms from the onboard computer. Between the computer nearly overloading, the lander running low on fuel, and an inconvenient crater in the landing area, the astronauts—and everyone watching—had a *lot* to worry about. The intensity was heightened since all these developments could only be heard, and not seen, back on Earth. All that was on the TV screen was a still picture of the *Eagle* lander and a clock counting down to the expected time of landing.

The audio of Buzz Aldrin talking to NASA's ground control while they figured out if the alarms were serious or not wasn't necessarily calming either. Like most people watching, Phil had little idea of exactly what was happening. It was mostly astronaut jargon and numbers:

"1202 alarm."

"We are go on that alarm."

"Five-and-a-half down."

"Nine forward..."

For all some viewers knew, these could have been signals of grave danger. In fact, they kind of were.

But there was one calming thought in Phil's head: Neil and Buzz were chosen for a reason. They were the greatest pilots on Earth. They could fly anything. Calming ideas of the astronauts' competence aside, the tension persisted.

"Contact light. Engine stop."

Phil knew immediately what had happened. They would only purposely stop the engine for one reason.

They were on the Moon.

At 9:17 p.m. Salisbury time, humans touched down on the Moon for the very first time. Phil breathed a heavy sigh of relief. The preceding tension was quickly replaced as a sense of excitement washed over him, realizing the astronauts were safe and that he would soon be able to watch them walk around on the surface of another world.

Granted, "soon" is a relative term. For *six hours* no one would leave the lander. The only activity was Neil and Buzz communicating with mission control. First thing on their to-do list was to make sure *Eagle* hadn't suffered any damage during landing. Once they were confident their ship was okay, they could settle in without risk of being stranded on the Moon with a broken spacecraft.

By about 11:00 p.m., Phil's parents were ready to go to bed. Lights around the house clicked off and Phil turned the volume down just a touch. He wasn't going to bed. He was very content to sit alone in the darkness of his family's living room, accompanied by only the glow of the TV. For Neil and Buzz, it was darkness and glow too—the darkness of space that surrounded them paired with the glow of the Moon's sunlit gray surface.

The BBC announcers began to talk about where it seemed the astronauts had landed. Apparently, the landing was quite close to where they expected, but no one knew exactly where for sure. Phil pulled out his scrapbooks to find maps he had collected from astronomy magazines over the years. For the first time he could look at these maps, as he had a hundred times before, and think, "They're actually there."

Another two hours passed as the news anchors and space experts talked about the ramifications of what had just happened. The coverage also reviewed the Space Race up until that moment. Sputnik, Laika, Yuri Gagarin, Alan Shepard, Valentina Tereshkova, Surveyor 3...it had all led to this. Phil, though, was particularly interested when the experts went over what the astronauts might be experiencing. The doctors at NASA who were monitoring Neil and Buzz's heartbeats from Earth confirmed to the news stations that the astronauts were doing just fine. Everything was going according to plan. These two guys were *on the Moon*, and somehow their hearts, though with briefly accelerated pulses, were normal.

Audio from the moment of landing was replayed more than a few times: "Houston, Tranquility Base here. The *Eagle* has landed." While a few news updates interrupted the BBC's coverage, they were committed entirely to covering the moon landing—save a brief discussion of some football scores from earlier that day.

This was a hurry-up-and-wait period for both Phil and the astronauts. It was now past midnight, and turning to the early hours of July 21. Most of Salisbury was asleep. Phil was not.

By the light of the TV, he was looking at maps of the region where the astronauts had just landed. His imagination conjured up views of what they might see when they stepped outside. There were photos taken by robotic landers that showed what the surface of the moon was like, but they were very limited. Those pictures were from one angle, at one time. In a little while, the astronauts would have cameras sending live footage right to Phil (and of course tens of millions of other people).

More time passed. The house started to feel very quiet. Noise from outside stopped. There was no more traffic at this hour.

Finally, around 3:30 a.m. Salisbury time, Patrick Moore announced that the astronauts were preparing to leave the spacecraft. As the experts on the BBC panel were discussing the astronaut's preparation for the EVA, the screen suddenly changed.

It was the Moon. There was live television beaming into Phil's house *from the Moon*.

The EVA was just beginning. Neil had opened the hatch to take a peek. Once he stepped onto the "porch" of the lander (the top of the ladder), he pulled a lever that deployed the Modular Equipment Storage Assembly, or MESA. The MESA was used to store things they needed on the surface, including the TV camera mounted on a deployable arm— basically an old school version of a selfie stick. Once the camera swung down, it turned on and began sending the signal back to Earth.

For Phil, the evening had now gone from the teen adventure of staying up all night, to a great voyage of exploration. The grainy TV signal from the moon wasn't exactly high definition; it was hard to even make out what was happening. But once Phil got a handle on things, he could see Neil Armstrong, a man about his father's age, standing at the top of a ladder just a few steps away from the moon.

Armstrong stepped down the ladder, stood on the footpad of the lander, and as Phil watched from his basement in Salisbury, the astronaut took "one small step" at 3:56 in the morning.

Phil vaguely recalled learning about Yuri Gagarin's first spaceflight, and Sputnik before that, but those were blurry childhood memories. Now he was watching *live* something that had never happened in all of human history, creating memories he knew he would have for a lifetime.

While he was experiencing this moment in tandem with millions of others around the globe, it was also a moment he shared alone with Neil and Buzz. Thinking about this over fifty years later, the Apollo 11 experience was a completely different experience than what we would expect today. Phil didn't have a group chat going with his friends during the mission. He wasn't refreshing Twitter to see the most recent information or perspectives on the event. He had no idea how other people were experiencing this or what their opinions were. Frankly, those things didn't matter. For Phil, and a lot of people watching Apollo 11, this was a personal journey to the moon.

Phil tried to make out as much as he could about where the astronauts had landed, a region called Mare Tranquillitatis. What was it like to be there? What was the landscape? Did it feel empty? Were there mountains on the horizon? Was the sun blinding? Could you see the bottoms of craters? Would the camera show things he could see on his maps? He wanted to know everything. And more than anything he wanted a map of where they were and what they could see. To Phil, that's what made exploration so special—the opportunity to see things no humans had ever seen and recreate them in maps and images that become part of history.

For the next three hours, Phil watched as the astronauts tried their best to describe their experience. From kicking up dust to collecting rocks, Neil and Buzz endearingly did many of the same things kids do when they're at the beach.

While Apollo 11's primary objective was to land rather than specifically do any science, once on the Moon, the astronauts were able to deploy scientific devices to, among other things, measure the Moon's seismic activity and the solar wind.

And just like that, after what felt like only a few minutes, at 6:11 a.m. the Moonwalk was over. Neil Armstrong closed the hatch behind him and the astronauts got ready for another first...humans were about to have the first slumber party on the moon.

Phil however was not moving from that chair in the living room. The glow of the television set started to be met by the early signs of dawn, the sky turning from dark to light.

The BBC began replays: the *Eagle*'s descent, the landing, the first step, and the EVA. Phil walked around the house a bit, grabbed some breakfast, and lived it all over again.

Around 9:00 a.m. replay coverage ended. As his parents went off to work, Phil had a new mission: collect as much stuff about the landing as possible.

•••

By the time of the morning rush, stores were already selling newspapers that had early information about the landing. A few of them had pictures taken directly from the television feed. One even had a rough transcript of what the astronauts said as they landed and walked around the Moon. Phil collected every single newspaper he could afford. The pictures weren't very nice, some of the initial information was inaccurate, and the transcripts weren't even close to perfect. But it didn't matter—Phil's scrapbook library was getting some major additions. It wouldn't be until a few days later, when the astronauts and their cameras were back on Earth, that the newspapers started printing nice big color pictures from the mission.

In the days following the landing, Phil heard people talking about the mission. A lot of people were excited, and he was happy that his town wasn't as in the dark about the Moon landing as he first thought. At the same time, he was surprised to hear some weird ideas about what had just happened. For instance, he heard some people saying that Apollo 11's landing ruined the romantic illusion of the Moon, since now it was just a gray ball of dirt. This wasn't a very accurate interpretation as far as Phil was concerned. People had known for a long time that the Moon was a big dusty rock. For him it was now a big dusty rock that we had walked on The Moon landing was an adventure like no other. It was an Arthur C. Clarke novel come to life.

As he heard more and more people talking about the Moon landing, Phil also began to learn about the political context of the Space Age. A lot of newspaper articles talked about how the Americans had won a race by beating the Russians to the Moon. This too seemed silly to Phil. He didn't care which flag got there first. To him, *humankind* had been on the Moon, and that was all that mattered.

The weeks went by, and magazines started to come out with gorgeous glossy pictures taken by the astronauts on the Moon's surface. Phil, of course, added these to his collection like a beaver amassing material for a dam. Before he knew it, news coverage shifted to the preparations for Apollo 12. He was still going through the treasure trove of images and articles from Apollo 11, and it was already time to do it all again!

With each additional Apollo mission, Phil became more and more interested in exactly where on the Moon the astronauts were landing and what kind of landscape they would see. After Apollo 11, Phil purchased *The Times Atlas of the Moon*. As he looked through the maps, he couldn't help but put little marks where the next missions were expected to land. Then after each Apollo mission, he added a circle around the mark to signify success. He also added the landing sites for all the Russian and American robotic probes that had reached the moon before Apollo.

As more successful Moon landings took place, more pictures from the surface of the moon were printed in magazines and newspapers. The detail with which Phil marked up the images he came across dramatically increased. If he saw pictures of a crater taken by the Apollo 14 lander as it descended, he would go to his *Times Atlas* and try to figure out exactly which crater it was. For the rest of the Apollo program, Phil spent much of his time building his own detailed atlas of the moon.

Eventually newspapers and magazines weren't enough. To meet his growing interest in more accurate mapping, he needed better pictures. So, Phil went directly to the sources. He wrote to both NASA and the Soviet Embassy in London asking for pictures of the moon. To his astonishment, both space agencies sent packages in return filled with large color images from the Apollo, Ranger, and Luna missions.

With all these various sources, Phil's hobby of marking up Moon maps was starting to match atlases made by professional magazines. Eventually, even the images directly from NASA and the Russians felt incomplete. Phil asked NASA for a catalogue of their lunar imagery so that he could identify what pictures might exist from certain areas, and which ones he might want for his collection. At the same time the Soviet embassy in London was sending him more detailed images from their lunar landers. The Russians even mailed him prints that he was allowed to make copies of before returning. Here was a seventeen-year-old English space enthusiast, in the middle of the Cold War, exchanging mail with the Soviets!

As he began university, Phil not only remembered the Apollo landings, but also an exhibition at the Geological Museum in London right around the time of Apollo 8, back when he first caught the space bug. The museum had printed huge pictures and maps of the Moon. There were models of what the surface might look like, and mockups of the Apollo spacecraft.

That day at the museum, Phil learned about how rocks tell us about the past. He learned that for a geologist, what a rock was made of explains how it was formed. Naturally, this meant that rocks from another world could tell us about how those worlds came to be. In a way, rocks could tell us about the *entire solar system* and its origins. As he prepared for his university education, he knew there was only one job he wanted.

It was a job that he hoped would allow him to keep up with the Space Age: He wanted to become a geologist.

•••

Decades later, Phil had become part of outer space in ways that he never thought he could as a seventeen-year-old secondary school student in Salisbury. His hobby of collecting pictures and maps of the moon turned into a real job. He became a geographer and professor of planetary science—studying the geology of other worlds. As an adult, Phil was now making a living doing essentially the same thing he had done since childhood. He kept up to date with space exploration, and created books and websites that collected, organized, and interpreted all the information he could find. Far from his atlases which, as a teenager, were just for him, lots of other scientists turned out to be interested in Phil's collections. Eventually he published definitive atlases of the exploration of the Moon and Mars. Now NASA was mailing *him* asking about maps.

Phil would go on to do even more. He developed new techniques for mapping asteroids, and worked closely with dozens of colleagues from NASA, the Russian space agency Roscosmos, and the wider space community. As a geography professor, he taught thousands of students about the Moon landings, planetary geology, and space exploration. In his years of teaching, Phil inspired many students to go on their own journey into space—just as the Apollo missions had inspired him, he was paying it forward a thousand-fold. As unimaginable as it had felt when he was young, through hard work and commitment to being a good person, Phil had forged his own path. He had become part of the global space community that once felt so far away.

Fifty years later, as July 20, 2019 rolled around, Phil was actually taking a day off from thinking about the Moon. He'd been thinking about the moon practically every day for five decades. Today, anniversary or not, he had already made plans to spend time with his family. But as he watched his granddaughter kick around sand and pick up rocks on the beaches of British Columbia, it was hard for him not to think about the Apollo astronauts. Who knows...maybe someday his granddaughter might be kicking up dust and picking up rocks on another world.

CHAPTER 5

Carme, Spain

Early in the history of our solar system, a violent collision of almost unimaginable scale occurred. It was an explosive rendezvous between a very young proto-Earth and rogue planetary embryo called Theia. Many planetary scientists and astronomers think this collision happened about 4.5 billion years ago. At that time, our sun was just starting to burn yellow, and the Solar System was a messy place. Small bits of rock and ice smashed into each other, sticking together and gathering like snowballs (what planetary scientists call "accretion"). As these dustballs grew to make larger and larger clumps of rock, they eventually formed planets, moons, and asteroids.

The early solar system was vastly different than the one we know today. Today, all of the planets and moons have pretty stable orbits, with only the occasional collision with an asteroid here and there. But back when the solar system had just formed, this was not the case. Planets and moons jockeyed for position. The gravity of the large planets pushed around the smaller ones, throwing asteroids around in all directions while they were at it. Eventually, around four billion years ago, the solar system settled down, and everything kept (mostly) to its lane.

When it collided with Earth, the smaller Theia was likely entirely destroyed. We don't have any way to know for sure. What we do know is that it left Earth radically changed. Large parts of the Earth melted entirely from the heat of the explosion and were forever changed. When it resolidified, Earth 2.0 emerged—and found itself with a new celestial neighbor.

Not all of the debris from the collision fell back to Earth. Instead, some of it began to orbit Earth like a ring. This mess of rubble eventually clumped together in the same snowball process that forms all large bodies in the solar system. The result of this particular accretion: Earth's new long-term partner, the Moon.

•••

A few billion years later, with humans populating much of the Earth, there was another violent collision. This one was not on a planetary scale, but rather, within the country of Spain.

The Spanish Civil War started in 1936 between those who supported the democratically elected government, and those who supported militaristic nationalism. The war lasted until 1939, and even played a role in the development of World War II. By the end, the dictator Francisco Franco had gained power and would rule over Spain for decades to come.

Thirty years after the Spanish Civil War, Barcelona, the capital of Spain, was the heart of Franco's regime. But for over a million Spaniards, Barcelona was also their home, as it was for eleven-year-old Carme. She came from a modest family, not overly wealthy, but living safely in a residential area of the ever-growing capital.

For an eleven-year-old in Franco's Spain, knowledge of the world beyond the country's borders was often limited. Even adults had difficulty finding information that wasn't censored or directly fabricated by the dictatorship. Because of this, the experience of the Cold War—and the Space Race—was very different in Spain than it was for other western European countries.

While much of eastern Europe lived under Soviet censorship, most western European countries had a relatively free and open press. Spain was the exception. The Franco regime strictly limited information about the outside world. Without a free press, news about the Space Race within Spain was uneven. Newscasts about Soviet accomplishments in space were few and far between, if they were mentioned at all.

Meanwhile, American news was much more common to come by in Spain because Franco was an ally of the United States. Both he and the US desperately sought to limit any influence from the Soviet Union on the Spanish population. The volatility of the Cold War saw dozens of revolutions around the world. Franco knew this and was constantly committing atrocities to suppress local communist and anarchist movements from overthrowing his dictatorship.

While Spain was indeed unique in western Europe due to its fascist regime, this experience of being exposed to select information about the Space Race was not at all uncommon. Around the world, countries' news programs were often edited based on who they were allied with in the Cold War. Even in Russia and America, the news was intentionally presented to make the other side look less appealing.

Despite all of this, news of the Space Age still fostered a sense of excitement among young Spanish children like Carme, unaware of the wider geopolitical circumstances.

Though there was no TV in Carme's home, she did see pictures of the Space Race in the newspapers. She would also sometimes go with her grandmother to see the nightly news on a TV mounted in the top corner of a local café, catching glimpses of a rocket launch or astronauts in space. While the rest of her family would often stay home, Carme lucked out in having a grandmother interested in space.

Watching the local café's small black-and-white TV, holding her grandmother's hand while they were in line to order, Carme took in what information she could. After, with juice in one woman's cup and coffee in the other's, they would sit together, looking up at the flashing images of the unfolding Space Age. Here Carme learned about the missions of Apollo: Apollo 8 to the Moon, Apollo 9 to test the LEM, and the Apollo 10 dress rehearsal; all played out in grainy black-and-white images.

•••

Having just completed primary school, Carme ventured to her aunt and uncle's house in Òrrius, a small village tucked amidst hills near Barcelona. Aside from the enjoyment of being away from the busy city, she was equally excited that her aunt and uncle's place had a TV. She knew that Apollo 11 would land on the Moon sometime that summer, and knowing she would be able to watch it live brought a smile to her face for weeks beforehand.

On July 16, Apollo 11 launched half a world away from Òrrius, Spain. Like all the moon landing missions, part of Apollo 11's flight plan was to fly one-and-a-half orbits around Earth before leaving for the Moon. Carme and the other kids heard people in the town saying that they might be able to see the light of the Apollo spacecraft flying through the sky. That afternoon, excited—but perhaps having not listened to the right people—she and her friends rushed outside to look. However, it was too early in Barcelona. Apollo launched in the afternoon, and by the time it left Earth orbit it was still only 6:30 p.m. in Spain. The sky was still too bright, and it would have been impossible to see any spacecraft in orbit.

Not being able to see the astronauts on their way to the Moon was a disappointment, but Carme was far from heartbroken. Her grandmother had come to Òrrius that week as well. Serendipitously, summer vacation had brought them together, with access to a TV, just in time to watch Apollo 11's journey. Their shared excitement for the Space Age would see its greatest high in just a few days.

Four nights later, Carme and her brother were fast asleep. As they dreamt their way into the early hours of the morning on July 21, a gentle hand on Carme's shoulder woke her up. It was her grandmother, waking her to let her know that the astronauts had successfully landed and were about to walk on the Moon. In the darkness of her room, Carme could hear the joy in her grandmother's whispered voice.

It was 4:00 a.m. Carme hadn't woken up this early too many times in her life. When she had, it was usually because of nightmares or noise outside—never to get up and go watch TV. Realizing she should hurry, she leapt out of bed like a kid on Christmas morning. Quietly stepping past her brother's bed, she and her grandmother made their way down the hallway and into the living room. Coming toward the end of the hallway, Carme could hear faint voices and see a dull glow. The television was already on; her grandmother had been watching through the night.

Just as Carme entered the room, the black-and-white screen turned bright with a mysterious view. It wasn't quite clear what was happening. A ghostly image was present, and there was some movement in the middle of the TV screen. Eventually the two of them could make out that this was the Lunar Module, and there was an astronaut standing on the side of it, descending a ladder.

Shortly after landing, the astronauts snapped this picture out the LEM window. Their first view of the lunar surface was baron, alien, and magnificent.

+ For All Humankind +

There was chatter in English, and then they could see the astronaut slowly stepping down the ladder rung by rung. The man's voice was muffled and cracked with static. It was an American voice, and the man seemed amazingly calm. As he descended each step, Carme woke up a little more, her eyes widening. They had initially sat down on the couch, but Carme crawled down onto the floor and leaned closer to the television. She quickly looked at her grandmother. Their excitement bounced off one another. The same excitement they shared all summer long had been building up anticipation for this moment. Together, lit only by the glow of the first images of men on the Moon, they watched history unfold.

As the astronauts took the first steps on the lunar surface, the light was continuing to creep above the horizon on Earth for Carme. With her young mind still rustling out of sleepiness, all she could think was that she had awakened to a different world. Today, the Sun was literally rising on a new era: a time when humans had walked on the Moon.

Through the thin drapes of her aunt and uncle's living room, the day began to set in. Other members of the family woke up and joined the viewing party. The astronauts were continuing to talk, carrying out scientific experiments and taking pictures. It was all new. Everything Carme saw had never been done before. Certainly she had never seen anything like it, and neither had anyone else. Even at age eleven, she knew she was taking in history, observing the greatest technical accomplishment ever achieved by humans.

Little did Carme know that Spain was playing a critical role in this human mission to the Moon at that very same time. Outside of Madrid, hours away from Carme, lay the Fresnedillas de la Oliva monitoring station. Two young engineers, Carlos González and José Manuel Grandela, had been listening diligently to the transmissions being sent to and from the astronauts as they prepared to set foot on the surface of the Moon. Because of the position of Earth relative to the Moon at the time of the landing, González and Grandela were actually the *first humans on the planet* to hear Neil Armstrong's iconic first words after landing on the surface of the Moon:

"Houston, Tranquility Base here. The *Eagle* has landed."

Despite Armstrong addressing the call to Houston, it was Spain that picked up the transmission to send to the rest of the world.

•••

When Carme felt the tides of change swelling in that early morning in 1969, she was right for a few reasons. Change would come to Spain, even if unrelated to Apollo. Six years after Apollo 11, the dictator Franco died, and the country began transitioning back to a healthy democracy. Around the same time, the Cold War began to calm. The Americans and Russians docked their spacecraft together in a symbol of newfound partnership in space exploration. News from the Soviet Union finally filtered out to the rest of the world, and citizens in Soviet countries were exposed to more information from the West.

Other changes, brought on directly by Apollo, would have been almost impossible for an eleven-year-old to foresee. Carme could have had no idea at that time how Apollo's legacy would play a role in the rest of her life. But indeed, decades later, her life became intertwined with the change she felt that bright summer morning.

The Apollo astronauts brought back twenty-two kilograms of lunar rocks and dust. Analysis of these rocks showed that the Apollo 11 landing site—and likely the rest of the Moon— was dominated by volcanic material that had been pushed around by old lava flows. In the billions of years since the Moon's formation, these lava flows had dried up and become solid rock. Microscopic analysis also told scientists that the moon samples had never been exposed to water, meaning the Moon was a dry place...and had been so for a *very* long time. Finally, the lunar dust had the very fine texture of something akin to baby powder. Scientists think this is from rocks on the surface of the Moon being smashed into smaller and smaller pieces over billions of years thanks to repeated asteroid impacts. This also meant that the Moon probably never had an atmosphere, which would have protected its surface from some of these impacts, as Earth's atmosphere does (but only if the asteroids are small enough!).

The samples from Apollo allowed humans to play detective with the Moon's past better than ever before. This information began to fill in *huge* gaps in our understanding of how the solar system must have formed and where the Moon came from. The story that emerged regarding the Moon's origin was one of chaos and violence—the collision between Earth and Theia, 4.5 billion years ago.

Carme now lives to tell and build upon this story. She is a professor of astronomy, working at a university and exploring the history of our universe. Oddly enough, she doesn't feel a direct relationship between the events in her aunt and uncle's living room that morning five decades ago and her job today. But as an astronomer, she knows better than anyone that the two are inseparably linked. She thinks that perhaps watching those first steps on the surface of the moon left a footprint on her subconscious, guiding her toward her love for astronomy in ways she might not have realized until later in life.

An explosion of awareness about our place in this universe was kicked off by the twenty-two kilograms of rock brought to Earth by Apollo 11. To go back in time and stop the Apollo missions from happening would be akin to stopping the toppling of the first domino in a chain reaction of knowledge about our universe that has been ongoing for five decades.

Fifty years later, this is what the Apollo program represents to Carme: Change. Change in the world, and change in our knowledge of our place in it. She is proud to be able to tell her students that she got to see the moment when it all began, as it happened. To have watched it with her grandmother that morning meant she, too, was part of the history of her chosen field, the history of astronomy and all of science.

But change comes in all kinds of forms, and Carme saw more than lunar science progress as she grew up. When she was a child, under the rule of Franco, women were expected to be caretakers of their homes. They were not allowed to become university professors. Thankfully, throughout the 1970s, women gained back their human right to education and self-determination.

That same decade, Spain also joined the ranks of spacefaring nations. The country launched its first satellite,

INTASAT, in 1974, and was a founding member of the European Space Agency a year later. Spain sent its first astronaut, Pedro Duque, into space in 1998. Pedro was only five years old at the time of the Moon landing—he was from the same generation of Spaniards as Carme, that watched, or listened, live as Apollo 11 happened.

Despite being an astronomer, when she looks to the sky Carme doesn't merely think about stars only from a scientific standpoint. Looking at the stars isn't reserved for studying astronomy. She also looks at the sky for the sake of pure enjoyment. For her, gazing at the sky is for curiosity, for dreaming, for the inspiration to try something new and change the world. She is able to separate the idea of looking at the stars to understand them from looking at them to simply bask in the wonder that is our universe.

As the astronauts walked on the moon, it may have been all business for them, but certainly they too had time to look up and wonder. After all, what got them there in the first place was wonder—wonder at whether humans could achieve the goal of reaching the moon; wonder at what they would encounter when they did get there; and wonder about both the past and the future. Carme's hope for the future is that humanity keeps this sense of wonder in mind, because you never know what kind of change it can ignite.

CHAPTER 6

Matias, Mexico

✦ For All Humankind ✦

The small wooden bookshelf in Matias' room housed a collection of science fiction classics, including Jules Verne, Ray Bradbury, and Robert Heinlein. Contained within their pages were the blueprints for a child's dreams of space travel. For most of human history, this was the only way space travel happened, in the imaginations of writers and on the page.

But dreams of humans traveling into space didn't stay on the pages of fiction. For Matias in particular, they became a reality within the first eleven years of his time on Earth.

In the 1960s, Matias watched the events of the Space Race closely. He followed along with the accomplishments of NASA's Mercury and Gemini programs. Reading about imaginary astronauts was replaced with the real deal. He felt a connection, as if these astronauts were going to space for him—and for everyone else on Earth. It felt as if they went up so that the rest of us down below could join them vicariously. Through the astronauts' descriptions of what it felt like to be in space, and the footage on the nightly news, everyone got to go along on the journey.

Matias was much more aware of the happenings in space than most children his age were, and indeed perhaps more than many adults in the small village of Santa Cecilia Tepetlapa, a town of about one thousand people on the southern outskirts of Mexico City. From here, much of what he managed to glean about the Space Race between the United States and the Soviet Union came through the radio. More tangibly, updates on the latest missions and which country had beaten which in each milestone were regularly splashed across the front pages of the local newspapers.

Mexicans like Matias felt strong ties to US space efforts. Their neighbors were no different from them except that the Americans lived north of the Río Bravo (known as the Rio Grande in the US) and Mexicans lived to the south. To Matias, they lived similar lives, ones that were relatable to each other. The Soviet Union, on the other hand, seemed vastly different. Living under an authoritarian regime, Russia was portrayed in the western media as an enemy that had to be defeated for the sake of democracy and freedom. Aided by American news media, that image permeated into Mexico.

At the same time however, while reading science fiction novels and browsing the newspapers, Matias paid little attention to where any of the satellites or space explorers were from. Ideological and national identity was far from his mind. All of the incremental achievements in spaceflight were human achievements, neither "American" nor "Russian." And as events escalated toward humans landing on the Moon, the excitement took over. Regardless of who made it to the Moon first, it would be a key moment in human history.

Television access was limited in Santa Cecilia Tepetlapa in 1969. There was only one television in the entire village—coincidentally in Matias' household. But the television schedule was restricted, with shows only airing between 6:00 p.m. and 10:00 p.m. If you were to turn on the TV at any other point in the day or night, you'd be greeted with the sight and sound of static.

In the days leading up to the landing, feelings of both nervousness and anticipation began to pulse through the village. The perception that this would be a momentous event, not just for the Americans, but for humanity in general, was widespread.

On the evening of July 21, 1969, watching Neil Armstrong and Buzz Aldrin on television taking their first steps on the surface of the Moon, it never occurred to Matias that he wasn't watching it in real time. The restricted television schedule meant his experience was delayed by hours. But this didn't lessen the impact in any way. In his mind, he was on the moon with Neil and Buzz, surrounded by the magnificent desolation described upon Aldrin's first view of the lunar landscape. It wasn't only the two astronauts on the Moon who were there, taking it all in at that moment—Matias, and everyone else watching back on Earth, was right there with them.

As well as one big communal adventure, it also felt to Matias like the marker of an epoch of accomplishment. Even at his young age, he recognized this was a moment that would divide all history books in the future: before and after the Moon landing. For humanity, 1969 would always be the point distinguishing pre-lunar and post-lunar landing.

The moon had suddenly become within reach, a part of our horizon. The solar system itself now felt a bit smaller. Human ingenuity had conquered the distance between Earth and the Moon—nearly 400,000 kilometers worth of science and engineering at work. From there, we could continue onward to wherever we put our minds.

Excitement and enthusiasm for Apollo 11 and the Moon did not wane quickly in Santa Cecilia Tepetlapa. Adults spoke of the event, filled with happiness, for years. Matias was able to relive the mission over and over again through the photographs printed in newspapers. Children built toy rockets out of anything they could find. While more privileged children had access to fancy packaged models, the lack of toy stores in Santa Cecilia Tepetlapa meant Matias and his friends used milk cartons and pieces of wood, hastily painted white to match the Saturn V. Their analogue for a Moon on which to land their rockets? Small rocks on the street—because large boulders would be too simple a task in their eyes!

Class projects in school also revolved around Apollo and the Moon for quite a while after the landing, spanning lunar geography and the history of lunar missions thus far. Many newspapers with articles and photos covering Apollo 11 were sacrificed to create posters and dioramas. The municipal government even teamed up with a Mexican colored-pencil manufacturer to host an art competition for kids. Matias won first place with his drawing of the *Eagle* lander on the surface of the Moon with the *Columbia* module overhead. The prize: a huge set of colored pencils so Matias could keep illustrating his dreams. His love of spaceflight was firmly entrenched, and before he knew it, Apollo would find its way into his life.

Despite being self-trained and self-educated, Matias' father worked with the Mexican Space Agency developing solid rocket fuels. It was an amazing time of innovation and opportunity. As the Apollo program continued, Matias' father secured a job at the Guaymas-Empalme ground station, operated by the Mexican Space Agency in support of the later Apollo missions. This ground station, along with others in the US, Nigeria, Kiribati, Bermuda, Australia, and the Canary Islands, were used during the Mercury and Gemini programs, reflecting the truly global effort behind Apollo. Through this, Mexico played its own key role in the Space Race, and connected millions of Mexicans, like Matias, personally to the Moon.

•••

Walking along the Camino de Santiago—a long-standing Catholic ritual built upon an even older pagan ritual dating back centuries—might seem like the farthest thing from humans walking on the Moon. The pilgrimage across southwestern Europe consists of a number of different routes, all leading to Santiago de Compostela, or the "Field of Stars." In Catholic tradition, the nearly one thousand-year-old Cathedral of Santiago de Compostela is the burial site of Saint Santiago. In English, he is known as Saint James, one of the twelve apostles of Jesus Christ and the patron saint of Spain. The cathedral spires tower over the surrounding buildings of the city, a dramatic amalgam of architectural styles spanning hundreds of years of construction and additions. Undoubtedly, it is a stunning sight for pilgrims to set their eyes on at the end of their long trek to this sacred place.

As an adult, Matias decided it was time to make his journey. He chose to travel along the Camino del Norte route, one of the lesser traveled paths to Santiago de Compostela. It stretches over eight hundred kilometers from the border of France and Spain along the Bay of Biscay toward the coast at the Atlantic Ocean. Pathways transition from dirt to stone, weaving among rolling hills, pastoral fields, and villages dating back hundreds of years. Churches and *albergues* (hostels) await the pilgrims along their journey, acting as waypoints along the path.

But for a space enthusiast like Matias, perhaps the most striking feature of this particular path for the pilgrimage is its alignment. At night, in the darkness away from heavily populated areas, the Milky Way streams along the sky immediately over the trail, leading travelers to their destination. The pilgrims on the Camino del Norte are quite literally guided by the galaxy to the Field of Stars.

This was something Matias' adventure shared with the Apollo program. As advanced as the guidance computer onboard Apollo spacecrafts was in the 1960s, the mission still relied on star charts for navigation. These charts are essentially maps of the sky. Since the stars don't change position over the course of a short mission, they can be used as guidance, just as sailors on Earth had done for thousands of years. Seemingly, wherever humans venture, the stars are there to help guide our way. Appropriately, Matias set out on his journey with a backpack adorned with a patch of the NASA logo and an Apollo 11 mission patch. He too was following the stars.

The pilgrimage was nothing short of epic. Walking over eight hundred kilometers in the span of a month to reach his destination, he accomplished a goal of which he had dreamt for years. Requiring five hours or more of walking per day, every day, is no small task for the human body, especially given the unpredictable conditions of the Galician coast. Morning fog could part to reveal the sun, or thicken to result in torrential downpours. To Matias, these were all just part of the experience.

This was, in some ways, Matias' own Moon shot. Some aspects of the journey were tests of endurance, others rewards for embarking upon such an effort in the first place. More than an accomplishment, the pilgrimage was its own reward, providing a spiritual retreat from modern life, and allowing for reflection, meditation, and mental serenity. For Matias, this was all encapsulated in one deeply humbling feeling: walking beneath the glow of the Milky Way, along the path so many had walked before him for so many centuries. From here he couldn't help but feel anything other than that he was just one small piece of an expansive universe.

•••

NASA had made the Moon landing look so flawless, so simple, that Matias expected humanity's steps beyond the Moon to happen quickly. Sadly, there was a feeling of stagnation after the Apollo program ended in 1972. Infrastructure supporting the Apollo missions at the Guaymas-Empalme ground station where his father had worked was decommissioned after Apollo 13. When the entire program came to an end after Apollo 17, he, like many, was left wondering when humans would return to the Moon. The years went by, and while humans continued to send robotic emissaries to the Moon and Mars, no human set foot on another celestial body after Astronaut Eugene Cernan, on December 15, 1972, left humanity's final boot prints on the Moon.

Matias followed in the footsteps of his father, who eventually became an economist. With the expertise of his own profession, he understands as well as anyone just how much of a challenge space travel is, whether it be engineering or, in the case of Project Apollo, economics. However, despite the slowdown in human exploration, he never lost his love for the space program. Matias' own enthusiasm for aerospace was passed along to his son, who is now an airline pilot with a deep love of space.

Sitting in his living room in Mexico City, reminiscing upon the Moon landing with his pilot son Saúl at his side, Matias thinks back on that era of bravado and accomplishment. "How adventurous we all are," he says, in Spanish, with a smile. The Moon landing was a singular moment in history that managed to bring the entire world together in a moment of hope, of camaraderie, of dreams for the future. This feeling is reflected in many of the goodwill messages etched into the disc left on the moon by Neil Armstrong and Buzz Aldrin.

It can be heartbreaking to think that this feeling didn't last for much of Earth's population. Matias sometimes wonders: What catalyst will be required to spur that feeling again, in a way that is sustained? What lesson should we learn fifty years after Apollo? "I hope that society will act more reasonably, and that we'll be able to better use our limited resources here on Earth."

This reflection was a common one in the aftermath of Apollo. Two of the most famous images of the entire program are not even of the Moon, but of Earth. Taken during Apollo 8, *Earthrise*, with the Earth juxtaposed to the barren moon in the foreground, was a wake-up call to humanity. Our planet was a paradise we needed to protect. A few missions later, Apollo 17's *Blue Marble* photograph of Earth would become an icon of the growing environmental movement in America and around the world. Alone in the photograph, against the darkness of space, Earth was a blue and green spaceship. It was another wake-up call. If humans want to survive, we had better get our ship in order. If the desolate photographs of the surface of the Moon taught us anything, it is how precious Earth is, and how lucky we are to have this hospitable place to call home.

Of all the images from the Apollo program, it was ths one that had the biggest impact. Contrasting the barren lunar surface against our beautiful home planet, Apollo 8's *Earthrise* image made clear there is no place like home.

All emotions considered, Matias always recalls the influence of Apollo with laughter and a smile, sharing it with his son, who reciprocates his father's joy. Like many of the Apollo generation, Matias feels gratitude at being able to share his memories of these great adventures with the world, and with his son. "In Spanish, we have a saying—*Recordar es vivir*. To remember is to live. Thank you for letting me relive this."

CHAPTER 7

Leena, India

The textbook definition of serenity can perhaps be found in the rural mountains of northeastern India. Here you'll find a lush green landscape blanketed in pine trees. Steep cliffs are sprinkled with stunning waterfalls cascading toward the valleys below. On a clear day, the peaks of the Himalayas can make an appearance on the northern horizon. Groves sacred to local tribes are sprinkled throughout the hills, the interior of which are riddled with dramatic limestone caverns.

Quiet serenity, however, was not something that was often found in Leena Das Gupta's home in the mountain town of Shillong. Rather, it was a bustling community—a joint family home of forty-five people. Grandparents, parents, uncles, aunts, cousins, multiple generations and branches of the family tree all under the same roof. Leena's home overflowed with music and merriment, conversation and camaraderie.

Tucked at the base of Shillong Peak, atop a broad plateau nearly a mile above sea level, is the hill station of Shillong. This high elevation creates a refuge from the scorching heat and humidity for which the bulk of the Indian subcontinent is infamous. Because of this, the British established the hill station during their colonial rule around the 1830s. The similarities in the landscape to that of the Scottish Highlands attracted Christian missionaries from Britain, who dubbed the area "Scotland of the East."

The British proceeded to build Christian churches and convents across the plateau. However, the region was not uninhabited when the British arrived. A local indigenous tribe—the women-led Khasi—already lived there, and were unsurprisingly none too happy about foreign invaders arriving on their land. After several revolts to oust the British, the Khasi were eventually defeated by the British army. However, their culture managed to persist.

During colonial rule in the 1850s, the British brought Dalit Sikhs from the opposite side of the country in Punjab to Shillong as manual laborers. Because of this, modern Shillong remains a tapestry of British, Khasi, and Sikh cultural influences. This fusion survived the Second World War, and continued after India finally gained its independence back in 1948.

The Space Race began shortly after the end of British rule in India. India's new place in the world, as a postcolonial state, would put it in the midst of the geopolitical friction of the Cold War. With the British driven out, India was working to build itself back up and establish its national identity. Along the way, there would be many hurdles. Two hundred years of colonial occupation had left its mark. The British had extracted much of the wealth and resources from India, leaving the country without much in the way of their own independent industry, electricity supply, or colleges. But given its geographic location and the context of the Cold

War, it wouldn't be long until another global superpower came knocking on the Himalayan doorstep.

In the 1950s, the Soviet Union began building strategic ties with India. Due to tensions with America, the Soviets did not want to import American movies, music, or culture into their country. Instead, they began importing Bollywood films. In a depressed postwar time, Bollywood was seen to bring a wave of cotton candy positivity and escapism to the Soviet population. Dubbed or subtitled into Russian for all to enjoy, Bollywood films became extremely popular across the USSR due to their high quality and beautiful music. To appeal even more to the Soviet audience, some Bollywood movies were even made based on beloved Russian stories. The Indian stars of these films, such as Raj Kapoor and Nargis Dutt, were revered in Russia the way Hollywood stars at the time like Audrey Hepburn or Cary Grant were in the United States.

But the relationship between the Soviet Union and India went much deeper than mere movies. The Soviets also provided India with economic and military assistance, building and financing multiple prestigious engineering colleges across the country. In general, many Indians at that point had a favorable view of Russia, which itself had shaken free of monarchy only half a century earlier. By the 1960s, both countries were trying to rebuild—one from war, the other from colonization. It was an odd, if not uneasy partnership, but many ordinary citizens on both sides felt it was a positive development in uncertain times.

In the eyes of some though, this view changed when the Soviets began launching dogs into space.

The Soviet Union launched multiple dogs into space during the early days of the Space Age, many of whom would become beloved celebrities, much like the later human cosmonauts. Sputnik 2, the second-ever mission into space, carried a passenger named Laika—a small terrier who became the first dog to travel into orbit. Tragically, Laika died during the mission. Five more missions carrying dogs into space launched between 1960 and 1966, with all but two of the dogs returning alive. The early days of the Space Age unfortunately saw terrible animal cruelty. (Luckily, we have progressed significantly in our treatment of animals in the space program, and these things are never done in space exploration today).

It was not the fate of the dogs, though, that raised the ire of many Indians. Leena's grandmother was of a generation of Indians who prayed to a Moon god in the evening and a sun god in the morning. The Moon was considered a sacred object. Even though the Russians were only launching dogs into the nearby space around Earth, there was a perception among some that they were launching dogs all the way to the Moon. This led to a view that the Soviets were desecrating a sacred place meant to be worshipped. The whole notion of a dog on the moon made Leena's grandmother rather upset. "What if the dog pees on the Moon? What happens then?!" her grandmother would ask.

•••

Nights in Shillong can get bitterly cold. The single radio in Leena's family home sat near the chimney, where the family would stay warm. Usually in the evenings, the women, men, and children alike, would gather around the radio and listen to local songs, singing and dancing.

But one night in July of 1969, there was no dancing.

All India Radio, India's most popular station, was not playing their usual musical fare. Instead, it was the news. The entire household was seated, glued to the broadcast, listening intently to every word. After a brief introduction, the newscaster's voice adjusted, as if what was going to be said might be heard many times over in the coming decades: Two American astronauts have landed safely on the Moon. The Apollo program is a success.

Watching the reactions of the adults around her, Leena saw what she could only interpret as shock. The adults seemed stunned. Human beings had made it to the Moon! When it sunk in, conversation emerged, questions and exclamations were exchanged. Eventually smiles and laughter dawned throughout the room, realizing the massive achievement by human scientists and engineers. The evening went from quiet anticipation to a lively crowd animatedly talking about what was next to come in this Space Age.

The reaction was certainly different from that of the snide remarks her grandmother made whenever there were reports of dogs being sent into space. The prevailing view regarding spaceflight among Leena's family was no longer one of desecration—instead, it was one of pride. Leena's grandmother, who had been angrily vocal at the Russians launching dogs into space, was quiet as men landed on the Moon. She was peaceful. The enormity of this moment in human history seemed to have set upon her.

For a child watching, it was a bit different. Leena's interpretation of the moment relied heavily on what she saw the adults around her doing. Leena knew it was a big moment, but seeing the way her family members were talking to each other with such excited voices made the entire experience feel that much more important, assuring that she would remember this moment for the rest of her life.

Newspapers were the only visual means of experiencing the Moon landing in Shillong. Household televisions were nowhere to be found at the time. Only newspapers hosted the images that would be discussed around the house. One of the first photographs to grace the front page of the local paper was the now iconic image of Buzz Aldrin's boot print in the lunar soil at Tranquility Base.

Armstrong snapped this picture of Buzz's boot print. Images like these became iconic, but they served a purpose too: Armstrong wanted a clear image of how compacted the lunar soil was when stepped on. Like fresh snow, it held the shape of boot prints in almost pristine shape.

For Leena, there was something about the boot print that felt wholly otherworldly. Was it the shape? The treads? The desolate...grayness of it all? At nine years old, standing in front of the newspaper stand at the market, Leena stared at this image with a sense of marvel and wonder. A single photograph of a single footprint, left by one of the first humans ever to set foot on another world in the entire history of our species. A literal and figurative giant leap in human history, captured within a single frame.

•••

Shillong modernized quickly in the years following the Moon landing. In the early 1970s it was designated the capital city of a newly formed state, Meghalaya. It became famous as a destination for music, shopping, and beautiful scenery. Across India, more and more schools and engineering colleges were established. Leena, who had been educated in a convent, went on to become a schoolteacher. While her career path was not influenced by the Moon landing, her profession allows her to appreciate the massive impact such things can have on the world, especially children.

While India had a good relationship with the Russians, the Americans were viewed as heroes by many for being the first to land on the Moon. It lit a spark for space travel all around the world, including India. While studying in school to become a teacher, Leena would often hear, "If they can do this, we can do this." And it was true—eventually India would indeed make its way to the heavens.

India's national space agency, the Indian Space Research Organization (ISRO), launched their first satellite into Earth orbit in 1075 from the Soviet Union. After that, India worked to develop rockets such as the Polar Satellite Launch Vehicle (PSLV). The PSLV is a world-class rocket now used by dozens of countries to launch satellites into space from India. Eventually, India chased their own lunar aspirations, launching their first mission to the Moon, Chandrayaan-1, in 2008. When it got there, Chandrayaan-1 helped to confirm the presence of water on the lunar surface, locked away in ice in permanently shadowed regions near the moon's south pole. Five years later, India ventured to the "Red Planet" with their Mars Orbiter mission, sending back a plethora of gorgeous color photographs. And India's second mission to the moon, Chandrayaan-2, launched just one day after the fiftieth anniversary of Neil Armstrong and Buzz Aldrin's landing.

•••

Looking back on the moment, fifty years later, Leena recalls the event with a smile. "Neil Armstrong, oh my god...he was a hero in those days!" she says with a bit of a schoolgirl giggle. Happily, she now has a son, Suman, a daughter, Indrani, two grandchildren, Adi and Rohit, along with a daughter- and son-in-law.

"Now that I'm old—I'm a grandmother now—we do talk about these things and how fast science has developed." She notes how her grandchildren have the world at their fingertips, able to access information about anything, anywhere, almost instantaneously. There is practically no delay in hearing about the events of the world, no waiting to hear what is happening out there. It is a striking difference compared to seeing the headlines each morning while wandering past a local newsstand, or gathering around the chimney for warmth while waiting for the news on the radio each evening. Life on Earth, and our world around us, changes in the blink of an eye.

But Neil and Buzz's footprints, and the boots themselves they left behind, lie in unchanged stillness on the Moon. In the absence of wind and water, the footprints persist all these years later with nothing to erase them away but the slow snow of microscopically small amounts of lunar dust ejected by asteroid impacts and transported by electromagnetic forces. Given millions of years, these layers of dust will slowly fill in the footprints. For those who witnessed those prints made, the dream remains that more human footprints will join them soon.

CHAPTER 8

Javad, Iran

سدق ری اطیا نکه ار هق ردب متمه

مرفسونن موددص قمه رتساز اردهک

Oh sacred bird, escort me with effort on my road,
as the way to my destination is long, and I am a new traveler.

It was surprising how green everything was. The pictures
hadn't done justice to how lush the plants in the yard actually
were. The house itself was tall but narrow, not a typical
home in North America, but a very common Persian home,
standing amongst others in downtown Shiraz. Outside
there was another feature common to homes in southwest
Iran: a shallow pool, not deep enough to swim around in,
but certainly useful on a hot day. Inside, the hallways were
narrow, gathering spaces were on the main floor, with the
living room and kitchen.

Peyman went up the narrow steps to the second floor where
the bedrooms were. At the top he saw multiple rooms with
their doors open. He wondered which one used to be his
father's. He had heard his dad talk about coming up here
many evenings to study or read. When he entered a small
brightly lit room near the back of the hallway, he knew he
had found it. This was where his father had fallen in love with
the stars.

•••

Historically, Iran has been a hotbed for stargazing. Between the eighth and fourteenth centuries, Persia (as the region was known then) was the center of a golden age of art, science, and philosophy. In this time, astronomy was a burgeoning field, and famous scientists like Al-Biruni (eleventh century) and Nasir al-Din Tusi (thirteenth century) ran some of the earliest observatories in human history. While there were dozens of amazing scientists at the time, Tusi alone was critical to humanity's understanding of the equinoxes, how the other planets moved, and Earth's position in the solar system. Not bad for one guy.

As a thirteen-year-old, Javad was just beginning to learn the basics of these subjects pioneered by Persian scientists. Introductory level algebra, trigonometry, and calculus filled his school schedule, as did, of course, astronomy. Most young teenagers in Shiraz knew the basics of astronomy— it was, after all, a part of their heritage. Like most kids, Javad learned to read the stars from an early age, and was curious from the time he was young about what might exist beyond Earth.

Missing out on the golden age of Persian science by about ten centuries was perhaps disappointing. A lot of the great truths of our universe had already been discovered, and having time to stare at the stars wasn't as common as it might have been centuries earlier. But Javad was living close enough to a different golden age, and he was sure to reap its benefits: the golden age of science fiction.

In the 1930s and 1940s, European and American science fiction authors such as John Campbell, Isaac Asimov, and Robert Heinlein, exploded in popularity. The demand for stories by these new writers also revitalized interest in older ones like Jules Verne and H.G. Wells.

Before humans had sent anything to space, our galactic neighborhood was a place of pure imagination. Tales of martian and lunar voyages and visitors combined actual scientific observations of the time with the minds of great writers and artists. The result was a generation of kids who spent most of their nights on different worlds.

The 1960s saw an influx of these science fiction classics translated into Farsi and made available in Iran. Just as they do today, teenagers had a tendency to fly through books at a dizzying pace, especially when they're on vacation from school. In the age before internet, TV, and cell phones, this was the world of teenagers—reading and getting lost in books (especially if there was bad weather).

A young Javad became immersed in the stories he collected. Jules Verne and Isaac Asimov particularly caught his attention. Even when his stories took place on Earth,

Verne described otherworldly places with enchanting detail. His books were often about the use of technology to explore places that felt alien, whether in space or not, such as the bottom of the ocean. But at the core was always science and adventure.

Asimov meanwhile could build an entire living world in but a few pages of a short story. Rarely writing full novels, Asimov's short stories took readers to space and back more effectively in a few pages than most could in a thousand-page novel. These writers didn't produce purely fantasy though. What made science fiction of the twentieth century so special was that it incorporated some of the most amazing discoveries of the era, describing state-of-the-art science along its way—and in some cases, predicting it. This grounding in reality would only feed Javad's and other teenagers' burgeoning interest in real-life space exploration.

The influx of western science fiction was not random. Iran and most of its citizens were considered to be aligned with the "western side" of the Cold War, with countries such as the United States and the United Kingdom. This meant Iranians were on the western side of the so-called "Iron Curtain" (the imaginary line between the east and west during the Cold War) and received a lot of American and British pop culture. As a result, a lot of Iranian citizens, like millions of people around the world, were largely unaware of events in the Soviet Union. That was the idea of the Iron Curtain—it was a metaphor for a dark curtain across the world where no light, no information, no ideas or cooperation, could pass in either direction.

But space, for the most part, was an exception. Being on the western side of the curtain, Iranians heard all about the developments of the American space program, as well as some snippets of Russian space accomplishments. In fact, one morning when he was just outside his house, Javad heard some of his family members talking to people about Laika, the first dog sent into space by the Russians.

Some of the people who heard that a dog had been sent into space were not happy at all. The reason? It was not because poor Laika never agreed to take such an uncomfortable and dangerous trip. Rather, dogs were seen as dirty animals to some in Iran, and it was sacrilege to send such a beast to the heavens, into God's domain. Javad didn't know what to make of this, but he was quite sure a cute little animal wasn't going to bother God.

•••

As the Apollo program was evolving, so too was Javad's education. Between 1963 and 1968, he progressed through middle school and secondary school into his final year. During those years Iranians became increasingly interested in the quest to send humans to the Moon. By this time,

TVs were slightly more common, and a generation of kids who grew up reading about human spaceflight in science fiction were ready to follow the real thing. In 1968, Apollo 7 galvanized the schoolchildren of Shiraz. Listening to the tall radio sitting on the living room floor, Javad and his family heard firsthand as NASA tested their Lunar Exploration Module (LEM) above Earth, setting the stage to go to the Moon.

The increasing rapidity of Apollo missions in 1968 resulted in an Apollo fever around the world. In Shiraz, this fever had some interesting side effects. Surrounding the buildup to Apollo, a rumor began to circulate around Javad's school that the Americans were accepting letters from people anywhere in the world who wanted to become an astronaut and go to the Moon. The story was that they would choose one person who wasn't an astronaut to join the Apollo 11 crew. Where this rumor came from, Javad never really knew, but the schoolchildren all talked about it in the hallway, perhaps exchanging ideas of how to convince the Americans to take them. One person who sent a letter was the school's janitor, who jokingly announced to the whole school that he would be the first Iranian in space. Javad listened to the radio a lot in those years; he never did hear about anyone from his school being blasted off into outer space.

The world was five months away from a new decade, and Javad had finished secondary school. His days were pretty relaxed at this point. It was a typical Monday in July, around five or six in the evening when he went downstairs to listen to the radio. The evening news would be on shortly, and the Apollo 11 mission had been happening all day (Shiraz was eight and a half hours ahead of the east coast of America). The radio was already on; his parents had been listening while preparing dinner. Javad sat on the midcentury wool couch in his living room and listened to the music of the local radio station.

By dinner time the weather had started to cool, the sun had begun its descent, but was still far from the horizon. July is typically the hottest month of the year in Shiraz, averaging nearly 40°C (104°F) highs each day. But one thing that is constant in the semi-arid region is a drastic change in temperature when the sun goes down. Passing the time, Javad tried guessing what was being made in the kitchen. Before he could pinpoint the aroma, however, he was interrupted. The music ended and the evening news was starting.

It started straight away with the Apollo program. The young-sounding announcer on the radio spoke excitedly, "Early this morning, American news media reported the successful landing and Moonwalk of Apollo 11."

Javad couldn't believe it. Starting to imagine what it might feel like to step on to the Moon, his thinking was again interrupted. The news played a short clip of English voices that begged his attention. It was muffled audio and Javad had no idea what they were saying. Balancing the competing emotions of curiosity and excitement, he looked at his parents, who had come in the room to hear the historic broadcast. Something about the voices sounded distant, almost like you could tell the transmission was coming from so far away. It was so much like what Javad had imagined when reading stories about humans on the Moon or Mars in science fiction, distant voices reporting sites previously unseen back home.

Javad realized how strange this really was. In so much of the science fiction he'd read, aliens were attempting to conquer Earth through violence. Now humans were the ones doing the conquering. But this was different. This wasn't about bringing violence or domination to the moon. The astronauts landed with the most peaceful of intentions: to explore, to learn for all of us, and to bring their stories back (along with pieces of the Moon itself). This kind of conquering was not about conquering each other, but about conquering challenges and exploring the universe though mastery of the little parts we've figured out. In this way,

+ For All Humankind +

it was intimately attached to all science that had come before, including the mastery of mathematics and physics developed by Persian scientists centuries earlier.

And that was just it. Javad found himself equally as proud of this achievement as he was of Iran's heritage of scientific excellence. He realized all of these accomplishments were part of the same global village, the same quest to understand the universe and our place in it. To use the tools of this understanding to connect us and make our lives better. Just as ancient trigonometry and algebra would allow us to build increasingly safe buildings and drinking-water systems, the Apollo program would lead to its own advancements for humanity, both in our daily lives and in the universal pursuit of knowledge.

Looking up from the radio, Javad could see his parents smiling. The sun was still up outside and the downtown chatter outside seemed to be lively, full of discussions and celebrations. It was a feeling of joy. Javad felt a unique, but still familiar, emotional state. The only thing he could compare it to was how it felt when his family would gather at the birth of a new child; the celebration of a new joy brought upon the world.

Humans landing on the moon had this same feeling—a new gift for everyone on Earth.

For so long a distant neighbor, when Armstrong and Aldrin snapped this picture on their way back home, the Moon, though having not changed at all, would forever look different to human eyes.

Having studied the progress of humanity coming out of the golden age of Persian science, Javad knew this accomplishment's impacts wouldn't be limited to any one part of the world. Just as it takes a village to raise a child, it took the global village to produce Apollo. The inspiration provided in return would lead to follow-up achievements everywhere.

•••

Primed by science fiction, people's guessing of what might come after Apollo was extravagant. The magazines of the time were filled with images of Moon bases, giant space stations, and missions to Mars, all projected to happen in the coming decades. While the drawings in the books sometimes looked like a dream, like something from the cover of an Isaac Asimov book, how could Javad, or anyone, doubt them? Only eight years after the first human entered space, humanity had just put people on the Moon. Who knew what could happen in another eight years?

Sadly for Javad, those paintings of Moon bases and Mars missions projected to become a reality by the 1980s and 1990s were not to be. The Apollo program had always been a big gamble—not only for the engineers and astronauts, but for the politicians supporting it. Project Apollo cost nearly $4 billion in 1969, which would be almost $100 billion today! This price tag meant it was not popular with everyone in America. Because of this, once it was successful a few times (six landings between 1969 and 1972), people felt it was time to bring it to an end. The goal had been accomplished and NASA's funding could be put elsewhere, such as on the

Voyager mission to the outer planets, the Viking Program slated for Mars, and the development of a reusable rocket system and spaceship—what would eventually become the Space Shuttle.

Though the Apollo program ended in 1972, Javad was far from done with space. As the years went by, and the Voyager probe flew past the gas giants of our outer solar system, he would keep up with the growing knowledge of our stellar neighborhood while raising a family with his wife. They would have a child of their own, and their families would gather to welcome the new boy into the world.

Just as millions of Persian fathers had done before him, Javad taught his son to read the stars, about the accomplishments of Iranian astronomers, and about the time humankind went to the Moon. Most of all though, he would teach his son the importance of science and its role in the global village. A child was to be guided by the wisdom that we are all in this universe together, seeking knowledge, and building upon what is given to us by those who came before.

•••

Looking at the calendar on his smartphone, Peyman couldn't believe it was already time for his father's visit, and that he would be hugging him in the airport baggage area in only a few hours. He had been excited for weeks, but he still needed to prep his father's room. Knowing his father Javad, Peyman figured it wouldn't be a bad idea to have some science or astronomy magazines on the nightstand. As an engineer on the Mars Curiosity Rover at NASA's Jet Propulsion Laboratory, Peyman certainly had enough space magazines lying around to put some in the spare room. He had also collected some glossy reports on the accomplishments of the Mars Rovers and martian science for his father to bring back home. Much cheaper than mailing them.

Peyman's job helping to operate a rover on Mars is proof of what Javad has always believed about new creations and this global village of ours. When you bring a child into this world you have no idea where it will go, what inspiration its creation will provide. But if there is the right village around it, that child's path can build upon what came before and lead anywhere.

EPILOGUE

Apollo was a global mission, the science and engineering of the Saturn V, LEM, and the CSM built upon centuries of knowledge, from the golden age of Persia to the European Enlightenment, from ancient rocketry in China to pioneering work in Tsarist Russia. The science fiction that inspired so many of the Apollo generation came from France, Russia, and England, among other places. Transmissions from the Apollo 11 crew were picked up and relayed all across the world, from Spain to Australia. Truly, the world as a whole was needed to inspire and create the reality of Apollo. And in return, as we've shown in the preceding chapters, Apollo reached back and inspired the world.

The stores of triumph and progress coming from Apollo mirrored millions of individual life stories. As humans have always been, we are creatures who face challenges and create change. The recognition that overcoming challenges is a collective experience is what this book has been all about. The shared endeavors of humanity bring to light one undeniable fact: everyone hopes and strives for a better future.

Unlike most mission patches, the Apollo 11 patch had few words on it. At their own request, the names of the astronauts were not included, as they wanted the patch to represent everyone involved in the mission. While the *Eagle* represented the home country of the astronauts, its talons bring olive branches, a symbol of peace for the prominent Earth in the background.

•••

The Apollo story did not end when Armstrong, Aldrin, and Collins splashed down on July 24, 1969. Seventeen weeks later, a new trio would make their way to the Moon. Pete Conrad, Dick Gordon, and Alan Bean repeated the accomplishment, this time pulling off a pinpoint landing in Oceanus Procellarum, the Ocean of Storms. Despite the much more hostile-sounding name compared to the Sea of

Tranquility landing site of Apollo 11, the follow-up mission was a total success. Attaining the capacity to land safely and with precision, the Apollo program was well on its way to placing LEM footpads in every region of the moon.

With the first one out of the way, NASA readied for a more precise mission, with a specific landing site and more focus on surface activities.

But the Apollo program did face challenges. Apollo 13 in 1970 was a near catastrophe. Its own claim to success was that no one was hurt after an explosion in the Command Module. All three astronauts—Jim Lovell, Jack Swigert, and Fred Haise—made it home safe. A lot has been written (and filmed) about Apollo 13, NASA's famous "successful failure."

Ironically, Apollo 13 is perhaps the second most famous Apollo mission after Apollo 11. As a result, people sometimes forget that there were four more, perfectly successful, Apollo missions in the two years after the near tragedy of Apollo 13.

With the tragedy of Apollo 1 still haunting the agency, NASA famously adopted "failure is not an option" as a mantra in getting Lovell, Swigert, and Haise back home safely. For all its magnificent achievements, the safe return of every space-faring crew is sometimes forgotten among Apollo lore.

In January of 1971, Apollo 14 would prove the stubbornness of humanity. After Apollo 13 was unable to land at its destination of the Fra Mauro highlands, the crew of Apollo

14 was undeterred and successfully landed in the vast hilly region. Apollo 13's objective was too important to pass up, so 14 needed to get it done. The reason for this comes down to basic lunar geography.

When you look at the Moon you can see two distinct features—light parts and dark parts. You don't need to be a geologist to interpret that. The dark areas are dried up seas, known as "mares," the Latin word for sea. But these were not seas of water.

Billions of years ago, these were seas of molten lava on the lunar surface caused by giant asteroids routinely hitting the moon. The bright parts are older, often mountainous, areas of the moon. These are called highlands. The importance of getting geological samples from both major lunar regions was paramount (keeping in mind that NASA couldn't assume later missions would be successful). The Apollo 14 mission was crewed by Edgar Mitchel, Stuart Roosa, and Alan Shepard, the second human to ever fly into space. Going to the moon wasn't all Shepard added to his resume on the trip. Amusingly, he also became the first lunar athlete when he hit two golf balls on the surface of the Moon.

Increasingly each Apollo mission intensified and focused more on science and less on symbolism. Pictured here is Alan Sheppard pulling equipment from a two-wheeled cart added to the LEM's cargo in order to help the astronauts move tools and experiments around.

For the Apollo 15 mission crewed by David Scott, James Irwin, and Al Worden seven months later, NASA got creative. The Moon is comparable to a rocky desert on Earth. There is sand-like material all over the ground, along with craters, valleys, boulders, and even mountains across its surface.

+ For All Humankind +

While the terrain wasn't necessarily impassable, it was slow moving for the astronauts in their large space suits, often carrying science experiments or tools. So, to help the astronauts move around, NASA sent them a car.

The Lunar Roving Vehicle (or LRV) was an electric buggy that allowed the astronauts to move more than ten times further than they could on foot. Packed inside the cargo hold of the LEM, it would fold out like a hideaway sofa bed and be left behind on the Moon when the mission was done. It looked like something from *Mario Kart*, and could fit both Moonwalking astronauts. It could reach a decent speed of about thirteen kilometers per hour, and had an onboard camera taking images every few seconds (and even some video).

A quirky legacy of the LRV was its role in proving people actually went to the Moon. For over forty years there have been people who don't believe it actually happened. Even today there are people who don't believe the footage, satellite images, sample analysis, or hundreds of thousands of eyewitnesses from the Apollo program. Good thing for reality: the LRVs had some speed to them. Footage of the LRVs driving on the Moon shows something truly otherworldly. The dust kicked up by the tires is thrown up like dirt from any dune buggy or dirt bike on Earth. But when it gets up off the lunar surface, it does something strange.

It dissipates outwards rather than falling straight down as it would on Earth.

What is happening is that the dust is not following the path we are familiar with on Earth, where we have a thick atmosphere and stronger gravity than the moon. The dust kicked up by the LRV is facing much less resistance in its flight from the lunar surface than it would on Earth. No one—not NASA, not the Soviet Union, not Hollywood—could have built an environment on Earth where dust moved that way. Not in 1971, not today. Decades after the landing, some of the last holdouts were finally convinced, humans had gone to the Moon, and six times at that.

Apollo was not a static program, it continually evolved. Nothing encapsulated that continual evolution as much as the inclusion of the LRV on Apollo missions 15, 16, and 17.

+ For All Humankind +

One of those additional missions was Apollo 16, the penultimate Moon landing mission. Apollo 16 landed deep in the highlands—the light areas we can see all the way from Earth. The astronauts on Apollo 16—John Young, Charlie Duke, and Ken Mattingly—expected to find small volcanoes littering their landing site. Instead, when Duke and Young went down to the surface, they realized right away that they were not among volcanic surface features at all, but rather ejecta from a massive asteroid impact millions—or billions—of years ago. In fact, the impact was so large that the debris they were looking at may have come from halfway across the moon.

By the end of 1972, the world was changing. Richard Nixon was the President of the United States and in charge of NASA, and for the time being, the relationship between Russia and the United States was becoming a bit friendlier. All throughout the Space Race, the controversial Apollo program had faced resistance (from, among others, many college-aged Americans, much of the Civil Rights movement, and major politicians) just as much as it had received support. America was ready to turn to other priorities. The writing was on the wall, Apollo 17 would be the last human mission to the lunar surface. However, before the program's end, the Moon had one last surprise up its sleeve for astronauts Gene Cernan, Jack Schmitt, and Ron Evans.

After being on the other end of the historic transmissions of Apollo 11, Charlie Duke went on his own grand voyage of exploration. Duke and Young walked on the dried remains of billion-year-old seas of lava, sampling rocks and craters that had formed in the eons since.

When Cernan and Schmitt went down the lunar surface, they discovered a few unexpected things. First, the area they landed in was supposed to be full of young ("young" according to geologists) rocky material. Instead, it was covered in ancient dust dug up from meters below the surface by, yet again, a giant asteroid impact from millions or billions of years ago. But this wasn't the only surprise. On the last visit it would have from its human neighbors for over five decades, the Moon showed its true colors. Literally.

When lava flows over rock it can create all sorts of interesting materials as the heat from the lava rapidly transforms the minerals inside the rocks. The result is completely new kinds of rock, called metamorphic rock. If you've played Minecraft, you'll know this well. Billions of years ago on the Moon, lava came into contact with rocks and melted them rapidly, turning them into glass (a common event on Earth). Apollo 17 astronaut Jack Schmitt stumbled upon these rocks and exclaimed, "It's all over, orange!" Due to higher than normal amounts of oxidized (rusty) magnesium, a huge patch of lunar soil is orange—the remnant of a volcanic event of the past.

By 1975, traveling to the moon was in the past. But this didn't mean the Apollo program was entirely done inspiring the world. In July that year, after three years away from space, NASA dusted off the Apollo hardware and launched a Saturn IB (the Saturn V's little sibling) into orbit. It wasn't going to the Moon. It wasn't even practicing going there. Onboard the CSM, Tom Stafford, Vance Brand, and Deke Slayton had a different mission, an entirely different kind of rendezvous to carry out.

While the Moon has certainly had color in the past, it wasn't necessarily expected that astronauts would find colors such as these on the surface.

Seven hours earlier, on the other side of the world, Alexei Leonov (the first person to conduct a spacewalk) and Valeri Kubasov launched from Kazakhstan in the Soviet Union and

+ For All Humankind +

entered Earth's orbit in their Soyuz spacecraft. They had the same date to make as the Apollo astronauts. The two spacecraft were to meet up.

In a show of international solidarity, both the American and the Russian governments had agreed that docking their craft together would be the ultimate feat to signify the cooperation needed in space exploration moving forward. After each crew conducted a few orbits, their spaceships met and docked. When the hatches between ships opened, the first thing to happen was a handshake and a space-hug between Deke Slayton and Alexei Leonov. The Cold War was calming. The "Space Race" was over, and the future of humanity in space would be cooperative. This is the true legacy of the Space Age: reaching out to new worlds, together, and making them more familiar.

The next time that humans set out to land on a new world for the first time, it'll have to lie beyond the Moon. And just as the destination will have changed, so too will the origin. In the fifty years since Apollo 11, Earth has transformed in seemingly unimaginable ways. We have an interconnected population of billions, twice as many as were alive in 1969. We are able to chat, share, and explore this world together instantaneously. We have a thriving scientific community that touches every continent. Space is now a realm of cooperation and science, a place where we learn to solve our problems on Earth while learning about the universe around us.

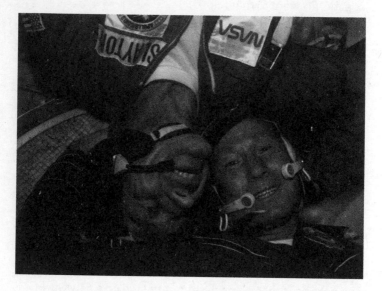

The end of an era, the international Space Age truly begins as Apollo ends. Scenes like this, of astronauts from different cultures embracing, are now common place. In 1976, it was groundbreaking.

•••

This new Earth, fifty years on from Apollo, has its own challenges. It may well be that the Moon shot of this generation is to address these. From reversing the causes and impacts of climate change and providing food for every

✦ For All Humankind ✦

person on Earth, to attaining true equality, there are no shortage of Moon shots in front of us. As we saw with Apollo, taking on these challenges can have unimaginably positive consequences. Pursuing any one of them (or others) will not only solve critical problems, but also inspire billions to lead better lives, contributing to the betterment of our global family.

Captured by Apollo 17 astronauts on their route to the Moon, this now iconic image remains salient nearly fifty years after it was taken. In this image is humanity's most precious spacecraft: in it sits the past, present, and future home of all humankind.

Apollo was a challenge. A big challenge. And that is perhaps the greatest lesson to learn from the stories in this book. Regardless of where our interviewees came from, the overarching theme was that of inspiration from seeing humanity overcome challenges. Crucially, none of the people we spoke with while writing this book were inspired because Apollo was about beating other people in a race, or making money. They were inspired by the great human accomplishment of setting a goal, being faced with challenges, and using the collective knowledge of the world to overcome them.

Challenges come in many forms: turning away from hate, adapting to change, inspiring the next generation, setting your own path, taking on epic journeys, pursuing an education, appreciating what you have, or raising a family. But all of these can be overcome through a mix of inspiration, hard work, serendipity, and love for our fellow humans who are there to help us on our way. Just like Project Apollo, and like everything humans have ever done, this is the message. There are always challenges. But when these challenges arise, either in our lives, or for the entire planet, no one is alone. Together, we can overcome any challenge in front of us, for our own betterment, and for all humankind.

DANNY BEDNAR

ACKNOWLEDGEMENTS

Our gratitude and appreciation goes out to the eight people who were gracious enough to share their stories with the world, and to their children, who acted as connections and translators to facilitate this project.

We would like to extend a heartfelt thank you to all 400,000+ people involved in the Apollo program over its eleven-year history, bringing the moon to all of us.

Special thanks to Nicole Mortillaro, Andrew Maynard, Julie Klinger, and Jeff Hopkins for helpful tips and words of encouragement as we embarked upon our own Moon shot of writing our first book.

+ For All Humankind +

ABOUT THE AUTHORS

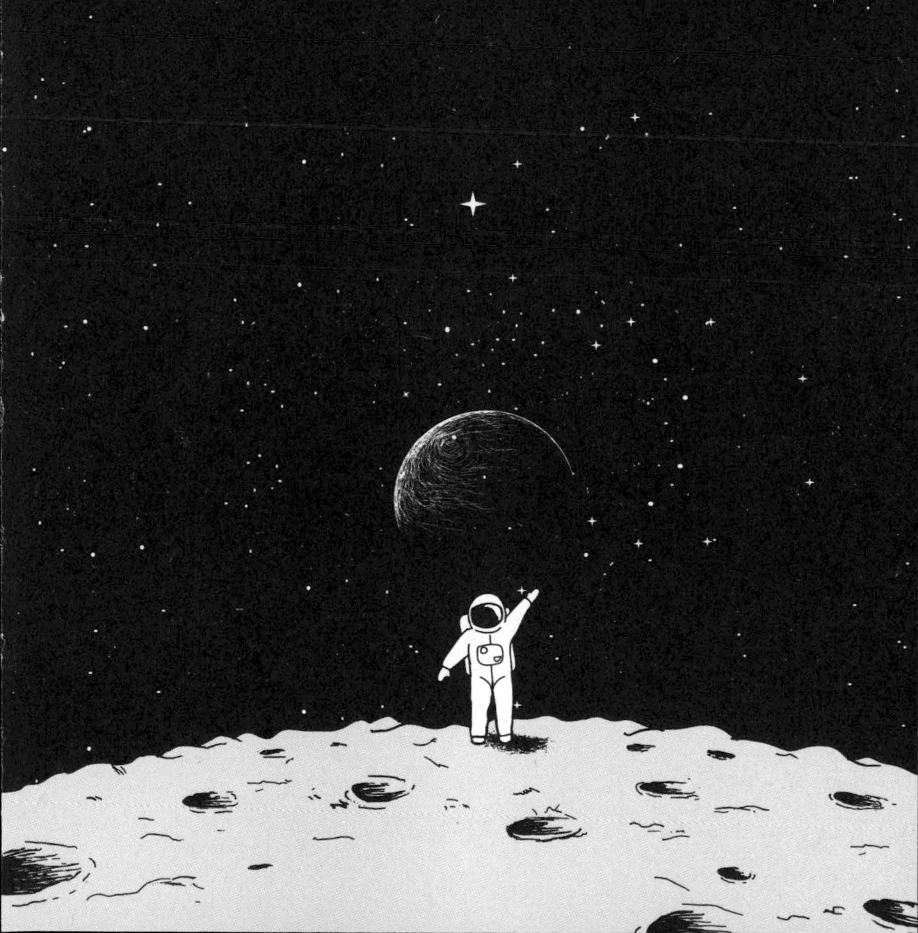

Dr. Tanya Harrison calls herself a "professional martian."
Over the past decade, she has worked as a scientist and
in mission operations on multiple NASA Mars missions,
including the Curiosity and Opportunity rovers. Her specialty
lies in geomorphology: the study of a planet's evolution
based on its surface features. Before Mars however,
Tanya had her head in the stars as an astronomer studying
the metal content of star clusters and recurring novae
systems. She holds a PhD in Geology from the University
of Western Ontario, a Masters in Earth and Environmental
Sciences from Wesleyan University, and a BSc in Astronomy
and Physics from the University of Washington. Tanya
is also an advocate for advancing the status of women
in science and for accessibility in the geosciences. You
can find her prolifically tweeting about the Red Planet—
and her experiences with both #WomenInSTEM and
#DisabledInSTEM—as @tanyaofmars.

Dr. Danny Bednar is a geographer of space and researcher with the Canadian Space Agency. His areas of interest include the use of satellites in the fight against climate change and the robotic exploration of our solar system. He holds a PhD in Geography, with a focus on climate change policy from Western University, Canada, where he has also been teaching about space exploration since 2012. Danny holds a Master of Arts in Geography with a specialization in Environment & Sustainability from Western University and received his Bachelor of Arts from the University of Winnipeg studying political philosophy and environmental science. Danny is passionate about mental health awareness and sharing his own experiences of living with severe depression and anxiety. He is also an advocate for the elimination of barriers for low-income students in education. In his spare time, Danny finds baroque music annoying and is passionate about providing a better life for all cats on Earth. He can be found on Twitter @DannyBednar.

Mango Publishing, established in 2014, publishes an eclectic list of books by diverse authors—both new and established voices—on topics ranging from business, personal growth, women's empowerment, LGBTQ studies, health, and spirituality to history, popular culture, time management, decluttering, lifestyle, mental wellness, aging, and sustainable living. We were recently named 2019's #1 fastest growing independent publisher by *Publishers Weekly*. Our success is driven by our main goal, which is to publish high quality books that will entertain readers as well as make a positive difference in their lives.

Our readers are our most important resource; we value your input, suggestions, and ideas. We'd love to hear from you—after all, we are publishing books for you!

Please stay in touch with us and follow us at:

Facebook: Mango Publishing
Twitter: @MangoPublishing
Instagram: @MangoPublishing
LinkedIn: Mango Publishing
Pinterest: Mango Publishing

Sign up for our newsletter at www.mango.bz and receive a free book!

Join us on Mango's journey to reinvent publishing, one book at a time.

✦ For All Humankind ✦